A Guide to SCILAB

With Applications

M. Affouf

Preface

This book is designed to show how to use the software package Scilab to help students solve problems from linear algebra, calculus, differential equations, and graphics.

This manual is organized into eight chapters. The first chapter introduces the general features of Scilab and its basic syntax. The second chapter contains the fundamental commands and Scilab functions of operations on matrices and applications to linear algebra. The third chapter introduces the programming features of Scilab. The fourth, fifth, and sixth chapters contain a more detailed look at most of the Scilab commands for graphing two and three-dimensional plots. Chapter eight is dedicated to basic data analysis. Seventh chapter contains applications to calculus and differential equations, and general mathematical problems.

I hope that this guide can contribute to developing mathematical software skills of Scilab and to better student understanding and performance in mathematical problem solving.

Finally, I would appreciate any comments, suggestions, and corrections which can be addressed to the email below.

M. Affouf

maffouf@kean.edu

Contents

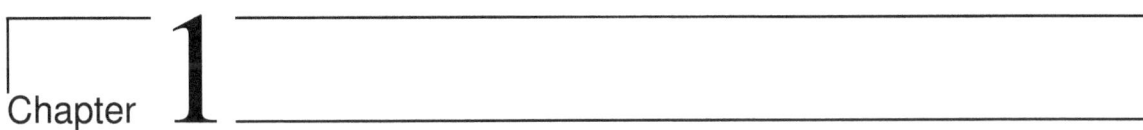

Chapter 1

Scilab Basics

1.1 Introduction

Scilab is a widely used and freely distributed mathematical software. Scilab is a high-level programming language for technical computing and interactive environment that integrates computation, visualization, and programming. It compromises three main components:

1. Scilab libraries of functions (procedures).
2. Libraries of FORTRAN and C routines. These libraries are mostly available from Netlib.
3. An interpreter which interacts with libraries.

Scilab has powerful capabilities to solve a linear system of equations and to perform advanced data manipulations. Scilab has powerful commands for two and higher-dimensional graphics. Scilab is a very efficient software for investigating and solving small and large data problems in areas of industry, commerce, research, and teaching.

Scilab can be used in a wide range of applications:

- Technical computing: Mathematical computation, analysis, visualization, and algorithm development
- Industrial research: Signal processing and communications, testing and measurements, image processing
- Teaching and Research in:
  - Matrix theory and linear algebra. differential equations, dynamical systems, and chaos theory.
  - Numerical analysis and scientific computing. Algorithms can be studied and evaluated in detail.
  - Statistics and probability concepts.
  - Engineering and scientific subjects, for example, physics, waves, and image processing. computational problems arising in economics, biology, and chemistry.

Scilab syntax is capable of manipulating matrices and all operations on matrix algebra. This allows users to solve many technical computing problems, especially those with matrix and vector formulations, in a fraction of the time it would take to write a program in scalar non-interactive languages.

Scilab has evolved for years and currently is developed by a team of dedicated researchers of Scilab Consortium which was created in 2003 by INRIA (the French national institute for research in computer science and control), the Scilab Consortium has joined the Digiteo Foundation in July 2008 and it includes industrial and academics, pursues the ambition of making Scilab the free reference in numerical computation.

Scilab features a family of add-on application-specific solutions called toolboxes. These modules (toolboxes) allow you to learn and apply specialized technology. These modules are made available to Scilab users directly from the Scilab console via a new feature named ATOMS (AuTomatic modules Management for Scilab).

1.2   Scilab System

The Scilab system consists of five main parts:
1. Development Environment. This is the set of tools and facilities that help you use Scilab functions and files. Many of these tools are graphical user interfaces. It includes the Scilab desktop and Command Window, a command history, an editor, and browsers for viewing help, the workspace, files, and the search path.
2. The Scilab Mathematical Function Library. This is a vast collection of computational algorithms ranging from elementary functions, like sum, sine, cosine, and complex arithmetic, to more sophisticated functions like matrix inverse, matrix eigenvalues.
3. The Scilab Language. This is a high-level matrix/array language with control flow statements, functions, data structures, input/output, and object-oriented programming features. It allows both "programming in the small" to rapidly create quick programs, and "programming in the large" to create large and complex application programs.
4. Graphics. Scilab has extensive facilities for displaying vectors and matrices as graphs, as well as annotating and printing these graphs. It includes high-level functions for two-dimensional and three-dimensional data visualization, image processing, animation, and presentation graphics. It also includes low-level functions that allow you to fully customize the appearance of graphics as well as to build complete graphical user interfaces on your Scilab applications.

1.3   Scilab Environment

We now discuss acquiring and installing Scilab. The starting point is the Scilab website at www.scilab.org. The homepage shows various icons and Scilab news. Click on the download link, then select the Linux, MacOS X, or Windows link. For example, I select the 64 version for Windows Scilab-6.1.1_x64.exe and download it to the computer. To install Scilab, click the downloaded Scilab-6.1.1_x64.exe file. The simplest procedure is to accept all default settings. To upgrade an installed Scilab program, you need to follow the downloading process described above. It is not a problem to have multiple Scilab versions on your computer. Scilab can be started by double-clicking the desktop shortcut icon.

The default Scilab desktop will then open on your screen (see Figure 1-1). As shown in the figure, the screen is divided into four main panels (windows). These are File listing in the current directory Figure

The default panels are
- *File Browser Window*: Shows the files in the current folder.
- *Command Window (Scilab Console)*: Main window, enters variables, runs programs. The first row includes file, Edit, Control, Applications, and ?. The most frequently used commands are listed in the second row of icons:
  1. **Launch SciNotes** to open the SciNotes Editor for writing scripts and functions.
  2. **Open file**

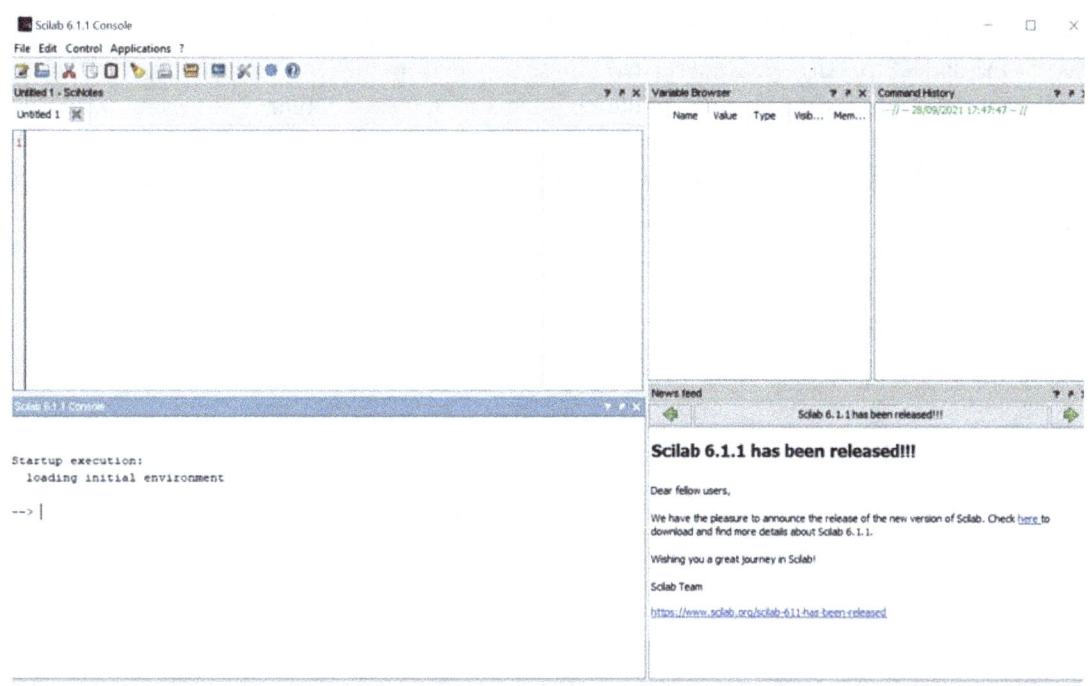

Figure 1.1: The default view of Scilab desktop

3. **Cut**
4. **Copy**
5. **Paste**
6. **Clear Console**, it will clear the Console viewing window, but then it will not delete the variables.
7. **Print**
8. **Module manager - ATOMS**, These are additional packages that enhance the Scilab capabilities. Some of these packages are not up to date.
9. **Xcos**, here you can find engineering modules and simulation packages.
10. **Scilab Preferences**, here you can change the fonts of SciNotes editor and the console fonts to make your typing easily readable. However, the default font types and colors are standard for comparisons and sharing.
11. **Scilab Demonstrations**, these are great tutorials, and it is a good idea to explore each one of them.
12. **Help Browser**, here you can find all Scilab commands and functions. You can explore it using word search or by topic. It is your reference to Scilab information and examples.

- *Variable Browser Window*: Provides information about the variables that are used.
- *Command History Window*: Logs commands entered in the Command Window.

Additionally, three more windows can be launched:

- *SciNotes Editor*: Creates and debugs script and function files.
- *Figure Window*: Contains output from graphic commands.
- *Help Window*: Provides help information.

All these separated windows can be docked into the Desktop by clicking the pointer on the dark blue stripe on the floating window and dragged to the desktop then release. You also can move the windows on the Desktop to different locations or arrangements following the same click, drag, and release of the dark strip. Also, you can separate these panels from the Desktop by click on the upward curved arrow. You can close any panel by clicking the x icon on the corner of the panel. It is a good idea to practice re-sizing the Desktop and dock the panels in the desired arrangement. If you close a panel, you can launch it by activating the Console and selecting the panel from the Applications group.

Console or Command Windows starts with prompt sign $--$> which is minus followed by greater. Click in it to activate the Console, then type $3+9$ after the prompt sign, and to execute this operation you should always press **Enter** key. The output will be

```
-->  3 + 9
 ans  =
    12.
```

General Scilab features

1. To type a command the cursor must be placed next to the prompt sign ($--$>).
2. Once an instruction is typed and the **Enter** key is pressed, the command is executed.
3. Several commands can be typed in the same line. This is done by typing a comma between the commands.
4. For long commands that is difficult to fit in one line, it can be continued to the next line by typing three periods . . . (called an ellipsis) and pressing the **Enter** key. The continuation of the command is then typed in the next line.
5. A previously typed command can be recalled to the command prompt with the up arrow key. When the command is displayed at the prompt sign, it can be modified if needed and then executed.
6. In Scilab the $=$ sign is called **assignment operator**. The syntax for assignment is

$$\text{Variable\_name} = \text{A value or an expression.}$$

For example,

```
--> x = 7
 x  =
    7.
--> x = 5 * x + 2
 x  =
    37.
```

7. Scilab is **case sensitive**, it requires exact match for variable names. For example, if you have a variable name x you can not refer to it as X. At the Command Window type define the constants: $a=4$ then press **Enter**, type $b=3$ then press **Enter**, type $c=5$ then press **Enter**, type $a+b+c$ then press **Enter**. Find the value $a+b$.
Typing $A+b$ and executing the line (**Enter**), leads to an error message:
!–error 4 Undefined variable: A. Why this massage and how to fix it?

```
--> a = 4
 a  =
```

```
      4.
--> b = 3
  b  =
      3.
--> c = 5
  c  =
      5.
--> a + b + c
  ans  =
      12.
--> A + b
    !--error 4
Undefined variable: A
```

8. Scilab accepts all kinds of data inputs: integer, real, complex, array, characters. No need to specify the type of data. All real numbers are used in double precision.

9. Scilab has *format* function to control the output of numeric values. The default format is 10 digits long. For example,

   • at the Command Window type sin(4) and **Enter**

   ```
   -->sin(4)
     ans  =
      - 0.7568025
   ```

   • to change the output to 16 digits, type *format (16);* sin(4) and **Enter**

   ```
   -->format(16)
   -->sin(4)
     ans  =
      - 0.7568024953079
   ```

   • to change the output to e-format numbers, type *format("e",10)*

   ```
   -->format("e",10)
   -->sin(4)
     ans  =
      - 7.568D-01
   ```

   • To change to default format (10 digits long), type *format(10)*
   • the maximum number of printed digits in Scilab is 25.
   • Use the Help Desk to see complete list of available formats.

10. In general, Scilab uses two types of extensions .sce is attached to script files and .sci is attached to function statements as will be explored later.

11. It is advisable that you create a new folder to house and save your files. For Scilab to access your files, you have to change the current directory to your newly created directory (folder). This can be accomplished on the Scilab (Current Directory) icon with the clicks of the mouse.

There are several general commands which are used frequently, they can be found as icons in the Scilab desktop window, or can be typed directly in the console:

General Commands

help	lists all available topics for help
clc	clears the console, does not affect the variables
clear	deletes all variables from the working directory and frees the memory
clf	clears figures
date	tells the date
exit	quits Scilab
quit	quits Scilab

1.4 Characters and Operators

Scilab has many types of operators such as arithmetic operators, conditional operators, logical operators, and others. It has also predefined characters which are reserved for specific tasks. For a complete list and information refer to the Scilab help. Here, we review some of these Scilab notations.

1.4.1 Special Characters

1. // **Two Forward Slashes** Are used to comment a line. Scilab ignores the text following // in a line.
2. No block comment operator in Scilab, yet!
3. $\begin{bmatrix} \cdots \end{bmatrix}$ **Brackets** are used to form vectors and matrices. For example, the vector X and the matrix M can be entered as follows

```
-->X = [1 2 4 8] //Enter
 X  =
      1.    2.    4.    8.

-->M = [1 2 3 4; 2 4 6 8; 1 3 5 9] //Enter
 M  =
      1.    2.    3.    4.
      2.    4.    6.    8.
      1.    3.    5.    9.
```

Brackets are also used to concatenate matrices and multiple outputs and also for arithmetic operations. These will be explored in the next chapter.

4. ( ) **Parentheses** are used to follow the order of arithmetic operations $3 - (7 - 9)$. They are used in the function statement $f(x, y)$ and to address a subscript in arrays $X(2)$ to call the second entry in the vector X.

5. , **Commas** are used for separation between matrix entries, between function arguments, and between instructions in a line. For example,

```
--> x = 2, y = 3, z = x+y
 x  =
      2.
 y  =
      3.
 z  =
      5.

-->X = [2, 4, 6] //Enter
 X  =
      2.    4.    6.
```

6. ; **Semicolons** are used after an expression or statement to suppress printing the output or the answer to console, and to to separate rows in a matrix. For example,

```
--> x = 2+4 //Enter
 x  =
      6.
```

```
-->y = 2+5; //Enter
--> A = [1, 2, 3, 4, 5, 6] //Enter
  A  =
      1.    2.    3.    4.    5.    6.
 -->B = [1, 2, 3; 4, 5, 6] //Enter
  B  =
      1.    2.    3.
      4.    5.    6.
```

7. : **Colons** are used to create vectors, array subscript, and for loop iterations. For example,

```
-->X=3:7 //Enter
  X  =
      3.    4.    5.    6.    7.
```

8. ... **Continuation** Two or more periods at the end of a line continue the current list and statements on the next line. This allows entering long statements requiring more than one line. For example,

```
--> 1 + 2 + 3 + 4 + ...
--> 5 + 6
  ans  =
      21.
--> 1 + 2 + 3 + 4 ...
--> - 5 + 6
  ans  =
      11.
```

9. $'$ **Apostrophe** is used to transpose matrices $B = A'$. For example,

```
--> X = [1 2 3]
  X  =
      1.    2.    3.
-->X'
  ans  =
      1.
      2.
      3.
```

10. **" " or ' '** **Two double quotation mark or two single quotation marks** are used to define string characters such as titles, colors and line marks.

1.4.2 Arithmetic Operators

1. + **Addition** between any numbers. Matrices of same sizes can be added and a number can be added to a matrix. For example,

```
-->2+7 //Enter
  ans  =
      9.
```

2. −   **Subtraction** between any numbers. Matrices of same sizes can be subtracted and a number can be subtracted from a matrix.

3. *   **Multiplication** of scalars or matrices. For example,

```
--> 3*8 //Enter
 ans  =
      24.
-->A=[1 2;3 4], B=[1 0; 2 5] //Enter
 A  =
      1.    2.
      3.    4.
 B  =
      1.    0.
      2.    5.
-->A*B //Enter
 ans  =
      5.    10.
     11.    20.
```

4. /   **Slash or matrix right division** for right division $B/A = B*inv(A)$. For example,

```
-->72/18 //Enter
 ans  =
      4.
```

5. \   **Backslash or matrix left division** for left division $A \backslash B = inv(A)*B$. For example,

```
--> 72\18 //Enter
 ans  =
      0.25
```

6. ∧ or **   **Power** for exponential operations on scalars and matrices. For example,

```
--> 16 ^(3/4)
 ans  =
      8.
-->(-2)**3
 ans  =
    - 8.
```

7. .*   **Matrix multiplication** element-by-element product. $X.*Y$ is equivalent to the element by element multiplication $X(i,j)*Y(i,j)$. For example,

```
-->X=[ 1 2 3 4], Y=[5 5 10 10], X.*Y //Enter
 X  =
      1.    2.    3.    4.
 Y  =
      5.    5.    10.    10.
 ans  =
      5.    10.    30.    40.
```

8. ./   **Matrix division** element-by-element quotient. $A./B$ is equivalent to $A(i,j)/B(i,j)$.

9. .∧   **Array power** is the element-by-element raising to power. For example,

```
-->X=[ 1 2 3 4], Z=X.^3 //Enter
 X  =
    1.    2.    3.    4.
 Z  =
    1.    8.    27.    64.
```

10. '   **Matrix transpose** For example,

```
--> A = [1 2 3; 4 5 6], B = A'
 A  =
    1.    2.    3.
    4.    5.    6.
 B  =
    1.    4.
    2.    5.
    3.    6.
```

1.4.3 Logical (Boolean) Operators

1. | **Logical OR** for scalars. The result T (true) or F (false). For example,

```
--> 3 > 5 | 3 >2
 ans  =
  T
--> 3 > 5 | 3 >9
 ans  =
  F
```

2. &   **Logical AND** for scalars. The result T (true) or F (false). For example,

```
--> 3 > 5 & 3 >2
 ans  =
  F
```

3. ∼   **Logical NOT** for scalars. The result T (true) or F (false). For example,

```
--> ~ 3 > 5
 ans  =
  T
```

4. $==, >, >=, <, <=, \sim=$ **Comparison Operators** Equal to, greater, greater or equal, less than less than or equal, not equal. For example,

```
--> 3 == 1 + 2
 ans  =
  T
```

1.4.4 Special Values

Scilab has few predefined special values (constants) which are preceded by the percent sign % .

ans   Most recent answer (variable). If you do not assign an output variable name to an expression, Scilab automatically stores the result in ans.

%eps   Floating-point relative accuracy. This is the tolerance Scilab uses in its calculations. Notice the effect of adding one and two epsilon on equality:

```
-->2 == 2 + %eps
 ans  =
  T
-->2 == 2 + 2*%eps
 ans  =
  F
```

%pi   is $\pi = 3.1415926535897...$

%e   is $e = 2.718181828...$

%i   Imaginary unit. For example,

```
-->(2+3*%i)*(2-3*%i)   //Enter
 ans  =
    13.
```

%inf   Infinity.

```
-->1/%inf
 ans  =
    0.
-->%inf + 5*%inf
 ans  =
    Inf
```

%nan   Not a Number, an invalid numeric value. Expressions like $\infty - \infty$ result in a %nan , as do arithmetic operations involving a %nan .

```
-->%inf - %inf
 ans  =
    Nan
```

%t or %T   is logical true

%f or %F   is logical false. For example,

```
-->%T | %F
 ans  =
  T
-->%T & %F
 ans  =
  F
```

1.5 Built-in Functions

Scilab has complete set of predefined elementary and specialized mathematical functions that can be located and reviewed in the on-line help documentation. We list the most common functions with examples to follow the tables.

Basic Functions

abs(x)	returns absolute value of x: $\lvert x \rvert$
sign(x)	returns the sign of x, 1 if $x > 0$, -1 if $x < 0$, and 0 if $x = 0$
sqrt(x)	returns the square root of x: $\sqrt{x}$
nthroot(x,n)	returns real nth root of a real number x. (If x is negative n must be an odd integer.): $\sqrt[n]{x}$
exp(x)	returns the exponential function of x: e^x
log(x)	returns the natural logarithm of x: $\ln(x)$
log10(x)	returns the logarithm with base 10 of x: $\log_{10}(x)$
log2(x)	returns the logarithm with base 2 of x: $\log_2(x)$

All Scilab functions are vectorized. You should use the Scilab command window to practice the following exercises:

1.5.1 Given the $x = \left[-2.3, -1.6, 3.1, 4, \sqrt{99} \right]$, calculate its absolute values and its signs.

Solution:

```
-->x = [ -2.3, -1.6, 3.1, 4, sqrt(99)]
 x  =
   - 2.3   - 1.6    3.1    4.    9.9498744
-->abs(x)
 ans  =
     2.3    1.6    3.1    4.    9.9498744
-->sign(x)
 ans  =
   - 1.   - 1.    1.    1.    1.
```
□

1.5.2 Which number is larger?
1. e^{π} or π^e
2. $\sqrt{2}$ or $\sqrt[3]{3}$
3. $\ln(e^3)$ or $\log(1000)$

Solution: *Calculate the values and compare:*

```
1. -->%e^%pi, %pi^%e
     ans  =
         23.140693
     ans  =
         22.459158
```

2. `-->sqrt(2), nthroot(3,3)`
```
    ans  =
        1.4142136
    ans  =
        1.4422496
```
3. `-->log(%e^3),  log10(1000)`
```
    ans  =
        3.
    ans  =
        3.
```

□

Number Theory Functions

factor(x)   returns the prime factorization of positive integer number x.
primes(x)   returns all prime numbers between 1 and x for positive integer number x.
factorial(x)   returns the factorials of positive integer number x.
gcd(p)   returns the greatest common factor of the components of positive integer vector p.
lcm(p)   returns the least common multiple of the components of positive integer vector p.
rat(x)   returns the approximation of the numerator and denominator of any real number x.

1.5.3 Answer the following:
1. What are the factors of Is $2^{23} - 1 = 8388607$?
2. Find the factorial of the vector $x = [3, 4, 5, 6, 7]$
3. What are the gcd and lcm of the numbers $12, 15, 24$?
4. Rationalize the numbers 0.5357143 and $\sqrt{2}$

Solution:

1. `-->factor(2^23-1)`
```
    ans  =
        47.      178481.
```
2. `--> x = [3, 4, 5, 6, 7]; factorial(x)`
```
    ans  =
        6.     24.     120.     720.     5040.
```
3. `-> x = [12, 15, 24]; gcd(x), lcm(x)`
```
    ans  =
    3
    ans  =
    120
```
4. *It is important to in save the output of rat function as a vector with two components: numerator and denominator as follows:*

```
--> [N, D] = rat(0.5357143)
 D  =
      28.
 N  =
      15.
--> [N2, D2] = rat(sqrt(2))
 D2  =
      985.
 N2  =
      1393.
```
☐

Trigonometric Functions

sin(x)	returns the sine of an angle x in radians: $\sin(x)$
sind(x)	returns the sine of an angle x in degrees: $\sin(x)$
asin(x)	returns the arcsine in radians of x: $\arcsin(x)$ that is $sin^{-1}(x)$
asind(x)	returns the arcsine in degrees of x: $\arcsin(x)$ that is $sin^{-1}(x)$
cos(x)	returns the cosine of an angle x in radians: $\cos(x)$
cosd(x)	returns the cosine of an angle x in degrees: $\cos(x)$
acos(x)	returns the arccosine in radians of x: $\arccos(x)$ that is $cos^{-1}(x)$
acosd(x)	returns the arccosine in degrees of x: $\arccos(x)$ that is $cos^{-1}(x)$
tan(x)	returns the tan of an angle x in radians: $\tan(x)$
tand(x)	returns the tan of an angle x in degrees: $\tan(x)$
atan(x)	returns the arctan in radians of x: $\arctan(x)$ that is $tan^{-1}(x)$
cotd(x)	returns the cotan of an angle x in degrees: $\cot(x)$
acot(x)	returns the arccot in radians of x: $arccot(x)$ that is $cot^{-1}(x)$
csc(x)	returns the cosecant of an angle x in radians: $\csc(x)$
cscd(x)	returns the cosecant of an angle x in degrees: $\csc(x)$
sec(x)	returns the secant of an angle x in radians: $\sec(x)$
secd(x)	returns the secant of an angle x in degrees: $\sec(x)$
sinh(x)	returns the sinh in x: $\sinh(x)$

For a complete list of all trigonometric and hyperbolic functions and their inverses refer to Scilab help documentation.

1.5.4 Calculate the following values
1. $\sin(\pi/6)$, $\cos(3\pi)$
2. $\sin(270°)$, $\tan(45°)$

Solution:

```
1. --> sin(%pi/6) , cos(3*%pi)
    ans  =
        0.5
    ans  =
      - 1.
```

```
2. --> sind(270), tand(45)
     ans  =
     - 1.
     ans  =
       1.
```

□

Rounding Functions

round(x)   returns the integer closest to x

fix(x)   returns the integer closest to x in the direction towards zero that is rounding up
               for negative numbers, and rounding down for positive numbers

floor(x)   returns the closest integer below x

ceil(x)   returns the closest integer above x

int(x)   returns the truncation of x, that is the integer before the decimal

modulo(x,y)   returns the remainder of the integer division $\dfrac{x}{y}$

1.5.5 Apply rounding functions to the vector $x = [-2.7, -1.3, 0.2, 1.6, 2.4]$

Solution:

```
--> x = [ -2.7, -1.3, 0.2, 1.6, 2.4];
--> [x; round(x); fix(x); floor(x); ceil(x)]
  ans  =
    - 2.7  - 1.3   0.2    1.6    2.4
    - 3.   - 1.    0.     2.     2.
    - 2.   - 1.    0.     1.     2.
    - 3.   - 2.    0.     1.     2.
    - 2.   - 1.    1.     2.     3.
```

□

1.5.6 Answer the following questions:
  1. Factor 182672
  2. is $2^{2^5} + 1$ prime?
  3. Find 20!
  4. Find the remainder from dividing $x = [1,2,3,4,5,6,7,8,9]$ by 3
  5. Find all prime numbers less than 22.

Solution:

```
1. --> factor(182672)
     ans  =
        2.    2.    2.    2.    7.    7.    233.
```

```
2. --> factor(2^(2^5) +1)
     ans  =
        641.     6700417.
3. -->format(25)
  --> factorial(20)
     ans  =
        2432902008176640000.
4. --> x = [1, 2, 3, 4, 5, 6, 7, 8, 9]
     x  =
        1.    2.    3.    4.    5.    6.    7.    8.    9.
  --> modulo(x,3)
     ans  =
        1.    2.    0.    1.    2.    0.    1.    2.    0.
5. --> primes(22)
     ans  =
        2.    3.    5.    7.    11.    13.    17.    19.
```

☐

Data Analysis Functions

max(x)	returns the largest value in each column of x
min(x)	returns the smallest value in each column of x
mean(x)	returns the average value in each column of x
median(x)	returns the median value of each column of x
sum(x)	returns the sum value of each column of x
product(x)	returns the product value of elements of each column of x
gsort(x)	returns the sorted values in ascending (increasing) order of each column of x

1.5.7 Given a vector $x = [51, 45, 55, 56, 65, 36, 47]$ and a matrix

$$A = \begin{bmatrix} 1 & 2 & 3 & 4 \\ 8 & 7 & 6 & 5 \\ -9 & 10 & -11 & 12 \end{bmatrix}$$

Answer the following questions:
1. Compute max(x), min(x), average of x, sort x in increasing and decreasing order
2. Find the sum of columns and rows of matrix A

Solution:

```
1. -->x = [51, 45, 55, 56, 65, 36, 47]
     x  =
        51.    45.    55.    56.    65.    36.    47.
  --> max(x) , min(x), mean(x), gsort(x), gsort(x,'lc','i')
     ans  =
        65.
     ans  =
```

```
      36.
   ans  =
      50.714286
   ans  =
      65.    56.    55.    51.    47.    45.    36.
   ans  =
      36.    45.    47.    51.    55.    56.    65.
2. --> A = [1, 2, 3, 4; 8, 7, 6, 5; -9, 10, -11, 12]
   A  =
      1.     2.     3.     4.
      8.     7.     6.     5.
    - 9.     10.  - 11.    12.
  --> sum(A)
   ans  =
      38.
  -->sum(A,1)
   ans  =
      0.    19.  - 2.    21.
  -->sum(A,2)
   ans  =
       10.
       26.
       2.
```

□

1.5.1 Scripts

Scilab is a collection of many programs that are either source codes or executable functions or user-defined scripts. A **script** is a sequence of Scilab instructions that are stored in a file (with extension .sce) and saved. The contents of a script can be displayed in the **SciNotes** editor. The script can be executed as a file or any part of the file. To create a script, click the Launch SciNotes icon or click Applications, then SciNotes. A new window will appear called the Editor. Type the sequence of statements, and save the file using File and then Save. Make sure that the **extension .sce** is on the filename (this should be the default). Save your file with a standard name so you can refer to it later. Make sure you created a folder to house all your Scilab work in a name such as work or Scilabwork. Write your name and the homework chapter with a date. Comments can be added highlighting the selection of lines with the mouse then click Format, then press Comment Selection, or comment with two forward slashes for each line. The following is a suggested example of a homework script saved as HW_Chapter1.sce

```
// Homework Chapter 1
// Affouf, M
// Date
/////////////////////
//1.6.1 Compute The following values
    sin(%pi/6)
    %e^2

////
//1.6.2 Compute the following values
    round(2.3), round(2.5)
```

1.6 Homework: Basics

Use Scilab to answer the following questions

1.6.1 Compute the following values
1. $sin(\pi/6)$
2. e^2
3. $\arccos(0.5)$
4. For $t = 36$ degree, find $\tan(t)$

1.6.2 Compute the following values
1. Round the numbers to the nearest integer: 2.3, 2.5, 2.7
2. The remainder of $\dfrac{48}{11}$
3. the prime factorization of 732
4. Rationalize the numbers 0.125 and π
5. compute 10!

1.6.3 Compute the following:
1. $6\dfrac{5}{17}$
2. $-\dfrac{3}{10} + 3^{-5}$
3. $-4.023^{4.023}$
4. Rationalize $\cos(\pi/5)$

1.6.4 Approximate the following numbers to 20 digits:
1. $\sqrt[5]{5}$
2. $\sqrt[6]{6}$

1.6.5 Find the digits of the periodic decimal representations of the following rational numbers:
1. $\dfrac{1}{3}$
2. $\dfrac{1}{7}$
3. $\dfrac{1}{13}$

1.6.6 Given that $a = 2$, $b = -3$, $c = 5$, compute the expressions
1. $a - b \cdot c$
2. $c + \dfrac{c - a \cdot b}{a - b - c}$
3. $a - \dfrac{b^3 - a \cdot c}{a \cdot c - b}$

1.6.7 Evaluate the expressions
1. 5^{3^2}
2. $(5^3)^2$
3. $5^{(3^2)}$

1.6.8 Evaluate the expressions
1. $\sqrt{3}, 2\pi, \dfrac{1}{e}$
2. $sin(\pi/3)$

3. $\sqrt{3}/2$

4. $\sin^2(\pi/3) + \cos^2(\pi/3)$

5. $\cos 45^0 - \dfrac{\sqrt{2}}{2}$

6. $\dfrac{2}{\pi} \cdot \arctan(\tan(\dfrac{\pi}{4}))$

1.6.9 Evaluate the expressions $e^{\pi i}$

1.6.10 Use logarithm properties to estimate the number 2016^{2017} in terms of powers of ten. For example $2^{15} = 32768$ is approximately 3.3×10^4.

1.6.11 Type the following into Scilab at the prompt command $-->$

1. $4 - 5*6$
2. $3.45*(-9.1)/3.27$
3. 3.05^3
4. $4^{(4^4)}$
5. $\sqrt{121}$
6. $\sin(pi/2)$
7. $12*4/2*12/3$
8. $3 \wedge (2+3)$
9. $(-3)^2 - \dfrac{5}{5+10}$
10. $\dfrac{4^{3-2} - 3}{8-3}$
11. $\sqrt{9 + 3^7}$

1.6.12 Use Scilab to define and compute the following

1. Define the array $x = \begin{bmatrix} 1.1 & 2.2 & 3.3 & 5 \end{bmatrix}$
2. Find the cosine of x
3. Subtract 1.1 from each element of x
4. Define array $y = \begin{bmatrix} 1 & 2 & 3 & 4 \end{bmatrix}$
5. Add the elements of x to the corresponding elements of y
6. Multiply each element of x by corresponding elements of y
7. Cube each element y
8. Define a matrix z of even numbers from 0 to 20
9. Use the linspace function to define an array v of all values between 3 and 12

1.6.13 Compute the area of the circle $A = \pi r^2$ with $r = \begin{bmatrix} 1 & 3 & \pi \end{bmatrix}$

1.6.14 Define a vector with values from 0 to 2π with increments of $\pi/20$

1.6.15 Define a vector with evenly spaced values from 5 to 100 with increments of 15

1.6.16 The distance of freely falling object is given by $d = \dfrac{1}{2} gt^2$ where $g = 9.8 m/s^2$. Construct a table of time versus distance for the first 10 seconds.

1.6.17 Use help menu to read how to use the functions

1. sin
2. tan
3. exp

4. ln
5. arctan
6. sqrt
7. abs

1.6.18 Create a vector x from -10 to 10 with an increment of 1
   1. Find the square root of the vector
   2. Find the absolute value of the vector
   3. Find e^x
   4. Find $\ln(x)$ and $\log_{10}(x)$
   5. Divide x by -3
   6. Find the remainder of x divided by 2

1.6.19 Using Scilab
   1. Factor 729. Use factor
   2. Approximate π as a rational number. Use rats
   3. Approximate e as a rational number
   4. Find 6! and 60!. Use factorial

1.6.20 Calculate the following
   1. $\cos(3\theta)$ for $\theta = \pi/4$
   2. $\sin(2\theta)$ for $0 \leq \theta \leq 2\pi$ with step size 0.2π
   3. Find $\cos^{-1}(.5)$

1.6.21 Using rand functions, compute the following
   1. Create 3×3 matrix of uniformly distributed random values
   2. Create 3×3 matrix of normally distributed random values

Chapter 2

Matrices

Matrices constitute the basic data type in Scilab. In Scilab, Matrix is a rectangular array of numbers. Scalars are considered one by one matrix. Vectors (columns or rows) are treated as matrices. Scilab has many tools to store numeric and character matrices, and to generate or import data tables. The best way to learn Scilab is to study how to handle matrices and the basic techniques of creating and manipulating matrices and their common applications.

Definition 1 A **matrix** is a rectangular table of numbers arranged in **rows and columns**. A matrix with m rows and n columns is referred as an $m \times n$ matrix denoted by A, the **elements** of A are called a_{ij}, where i indicates the row index and j the column index. For instance,

$$A = \begin{bmatrix} a_{11} & a_{12} & a_{13} & \cdots & a_{1n} \\ a_{21} & a_{22} & a_{23} & \cdots & a_{2n} \\ a_{31} & a_{32} & a_{33} & \cdots & a_{3n} \\ \vdots & \vdots & \vdots & & \vdots \\ a_{m1} & a_{m2} & a_{m3} & \cdots & a_{mn} \end{bmatrix}$$

- The **size** of a matrix is the number of rows m and the number of columns n.
- An $n \times n$ matrix is said to be **square**.
- A **row vector** is a matrix consists of only one row.
- A **column vector** is also a matrix of one column.
- A **scalar** is a matrix of size 1×1.

2.1 Generating Matrices

Matrices can be entered into Scilab in different ways:
- Type the elements of the matrix
- Use special built-in commands
- Load the matrix from external file

For example to enter the 2×3 matrix $A = \begin{bmatrix} 2 & 1 & 2 \\ 3 & 2 & 3 \end{bmatrix}$, we follow the steps:

1. Assign the variable A to **square brackets**: $A = [\quad]$
2. Enter the elements of the first row separated by blank spaces or commas and end the list with semicolon ; to indicate the end of the row: $A = [2 \quad 1 \quad 2; \quad]$

---

3. Enter the second row: $A = [2 \quad 1 \quad 2; \quad 3 \quad 2 \quad 3 \quad]$, the last row does not require ;

4. Press enter or run the line

As follows

```
--> A = [2 1 2;3 2 3]
 A   =
    2.    1.    2.
    3.    2.    3.
```

or by entering

```
--> A = [2, 1, 2; 3, 2, 3]
```

Row and column vectors are entered as follows

```
-->X = [1 2 3 4 5], Y = [ 1; 2; 3; 4]
 X   =
    1.    2.    3.    4.    5.
 Y   =
    1.
    2.
    3.
    4.
```

Scilab has a built-in function to generate famous matrices: Magic, Frank, and Hilbert. These matrices are studied in Linear Algebra courses.

2.1.1 Construct a Magic matrix of dimension 3 using Scilab.

Enter the matrix A by taking advantage of the pattern in the matrix

$$M = \begin{bmatrix} 8 & 1 & 6 \\ 3 & 5 & 7 \\ 4 & 9 & 2 \end{bmatrix}$$

We use the command

```
--> M = testmatrix('magi',3)
 M   =
    8.    1.    6.
    3.    5.    7.
    4.    9.    2.
```

Let us check why it is called magic?

```
--> sum(M,"c")  // column sums
 ans  =
    15.
    15.
    15.
--> sum(M,"r")  // row sums
 ans  =
```

```
     15.    15.    15.
--> diag(M)
 ans  =
    8.
    5.
    2.
--> sum(diag(M))
 ans  =
    15.
```

2.1.2 Construct 4 by 4 and 5 by 5 magic matrices and check their magic numbers (the sum of each row, column, and diagonal).

Hilbert matrix is a square matrix with entries being the unit fractions

$$H_{ij} = \frac{1}{i+j-1}$$

For example,

$$H = \begin{bmatrix} 1 & 1/2 & 1/3 \\ 1/2 & 1/3 & 1/4 \\ 1/3 & 1/4 & 1/5 \end{bmatrix}$$

2.1.3 Construct a Hilbert matrix of dimension 5 by 5 using Scilab.

```
--> Hi = testmatrix('hilb',4) // This generates the inverse matrix of Hilbert
 Hi  =
    16.    - 120.     240.    - 140.
  - 120.    1200.   - 2700.    1680.
    240.   - 2700.    6480.   - 4200.
  - 140.    1680.   - 4200.    2800.
-->format(6)
-->H = inv(Hi) // H is the Hilbert matrix
 H  =
    1.        0.5        0.333      0.25
    0.5       0.333      0.25       0.2
    0.333     0.25       0.2        0.167
    0.25      0.2        0.167      0.143
```

Hilbert matrices are examples of ill-conditioned matrices.

2.1.4 Enter the matrix A by taking advantage of the pattern in the matrix

$$A = \begin{bmatrix} 1/2 & 1/3 & 0 & 0 \\ 1/3 & 1/2 & 1/3 & 0 \\ 0 & 1/3 & 1/2 & 1/3 \\ 0 & 0 & 1/3 & 1/2 \end{bmatrix}$$

Type at the prompt sign

```
-->x=1/2;y=1/3;

--> A=[x y 0 0;y x y 0; 0 y x y;0 0 y x]
 A  =
    0.5         0.3333    0.        0.
    0.3333      0.5       0.3333    0.
    0.          0.3333    0.5       0.3333
    0.          0.        0.3333    0.5
```

2.1.5 Enter the Vandermonde matrix B by taking advantage of its pattern:

$$B = \begin{bmatrix} 1 & 1 & 1 & 1 \\ x & y & z & w \\ x^2 & y^2 & z^2 & w^2 \\ x^3 & y^3 & z^3 & w^3 \end{bmatrix}$$

Where $x = 2, y = 3, z = -2, w = -3$

Solution: *Construct numerical row R as follows*

```
-->x=2; y=3; z=-2; w=-3;
-->R=[x y z w];
-->B=[R.^0; R; R.^2; R.^3] // Note the dot power:element by element
 B  =
    1.     1.      1.     1.
    2.     3.    - 2.   - 3.
    4.     9.      4.     9.
    8.    27.    - 8.  - 27.
```

$\square$

2.1.1 Generating vectors

The direct way is to put the values that you want in the vector in square brackets, separated by either spaces or commas. For instance, both of these assignment statements create the same row vector v:

```
-->v = [1 2 3 5 8 13]
 v  =
    1.    2.    3.    5.    8.    13.
-->v = [1, 2, 3, 5, 8, 13]
 v  =
    1.    2.    3.    5.    8.    13.
```

2.1.6 Use the following US population table, to create a vector containing the population and save the vector as uspop.sod for later use.

Year	2010	2005	2000	1995	1990	1985	1980	1975
Population(Millions)	310	296	282	263	249	238	227	216

Solution: *We define the vector uspop and follow the procedure to save and load files:*

```
--> pop = [310, 296, 282, 263, 249, 238, 227, 216]
 pop  =
    310.    296.    282.    263.    249.    238.    227.    216.
-->save('C:\Users\M\Desktop\swork\uspop.sod', 'pop')
-->clear
-->pop
   !--error 4
Undefined variable: pop
-->load('uspop.sod','pop')
-->pop
  pop  =
    310.    296.    282.    263.    249.    238.    227.    216.
```
□

For Uniformly spaced vectors, Scilab has two commands: The **colon operator** : and the *linspace* function to generate vectors:

vector_name = a:b	returns a sequence of numbers starting with *a* and ending with *b* in steps of one.
vector_name = a:s:b	returns a sequence of numbers starting with *am* and ending with *b* in **steps of** *s*
vector_name = linspace(a, b)	returns a vector with 100 equally spaced values between *a* and *b*
vector_name = linspace(a, b, n)	returns a vector with *n* equally spaced values between *a* and *b*

Note that when we use the colon operator or the linspace function, we do not need to include brackets.

2.1.2 Dimensions

The **size** and **length** functions in Scilab are used to find array dimensions. The length function returns the number of elements in a vector. The size function returns the number of rows and columns in a matrix.

Examples

1. To create a vector running from 1 to 8:

```
--> X = 1:8
 X  =
    1.    2.    3.    4.    5.    6.    7.    8.
-->size(X)
  ans  =
```

```
           1.     8.
    -->length(X)
     ans   =
        8.
```

2. To create a vector of even numbers running from 0 to 12, Use an increment (step size of 2):

```
    --> X = 0:2:12
     X   =
        0.    2.    4.    6.    8.    10.    12.
    -->length(X)
     ans   =
        7.
```

3. Step size can be positive, fractional or negative:

```
    --> X = 0:0.2:1
     X   =
        0.    0.2    0.4    0.6    0.8    1.
    --> X = 14:-3:-5
     X   =
        14.    11.    8.    5.    2.   - 1.   - 4.
```

For example, to generate a vector *Z* with 9 uniform points between 0 and 1. Enter

```
-->Z=linspace(0,1,11)
 Z   =
    0.    0.1    0.2    0.3    0.4    0.5    0.6    0.7    0.8    0.9    1.
```

At the prompt sign enter the following commands

```
--> v2 = linspace(0,2,2); v5 = linspace(0,2,5);
--> v9 = linspace(0,2,9); v11 = linspace(0,2,11);
-->disp(v11,v9,v5,v2)
    0.    2.
    0.    0.5    1.    1.5    2.
    0.    0.25    0.5    0.75    1.    1.25    1.5    1.75    2.
    0.    0.2    0.4    0.6    0.8    1.    1.2    1.4    1.6    1.8    2.
```

2.1.3 Creating Column Vectors

To create column vectors using the colon operator or the linspace command

There is no direct way to use the colon operator or the linspace function to get a column vector. However, any row vector created using any of these methods can be **transposed** to get a column vector.

2.1.7 Create the vector

$$Year = \begin{bmatrix} 2004 \\ 2006 \\ 2008 \\ 2010 \\ 2014 \\ 2016 \end{bmatrix}$$

We use the commands

```
--> Y = 2004:2:2016
 Y  =
     2004.    2006.    2008.    2010.    2012.    2014.    2016.
--> Year = Y' // The prime is for Transpose
 Year  =
     2004.
     2006.
     2008.
     2010.
     2012.
     2014.
     2016.
```

2.1.4 Concatenating Matrices and Vectors

Matrices and vectors can be **concatenated** (combined) column-wise and row-wise as long as they match the column or row dimensions. To glue or stack matrices or vectors we use brackets. See the following examples.

2.1.8 Let

$$A = \begin{bmatrix} 10 & 20 & 30 \end{bmatrix} \qquad B = \begin{bmatrix} 1 & 1 \\ 2 & 8 \\ 3 & 27 \end{bmatrix} \qquad C = \begin{bmatrix} 1 & 2 & 3 \\ 4 & 5 & 6 \\ 7 & 8 & 9 \end{bmatrix}$$

Here are few variety of combinations

```
--> A =10:10:30
 A  =
     10.    20.    30.
--> B = [1 1; 2 8; 3 27]
 B  =
     1.    1.
     2.    8.
     3.    27.
--> C = [1 2 3; 4 5 6; 7 8 9]
 C  =
     1.    2.    3.
     4.    5.    6.
     7.    8.    9.
```

```
--> [A A]
 ans  =
     10.     20.     30.     10.     20.     30.
--> [A' B]
 ans  =
     10.     1.     1.
     20.     2.     8.
     30.     3.     27.
--> [B C]
 ans  =
     1.     1.     1.     2.     3.
     2.     8.     4.     5.     6.
     3.     27.     7.     8.     9.
--> [C ; B']
 ans  =

     1.     2.     3.
     4.     5.     6.
     7.     8.     9.
     1.     2.     3.
     1.     8.     27.
--> [C C; C C]
 ans  =

     1.     2.     3.     1.     2.     3.
     4.     5.     6.     4.     5.     6.
     7.     8.     9.     7.     8.     9.
     1.     2.     3.     1.     2.     3.
     4.     5.     6.     4.     5.     6.
     7.     8.     9.     7.     8.     9.
```

2.1.9 Construct a matrix that will hold the pair of vectors $x, -x$ where $x = [0,1,2,3,4,5]$.

```
-->x=0:5;
-->A=[ x ; -x]
 A  =
     0.     1.     2.     3.     4.     5.
     0.   - 1.   - 2.   - 3.   - 4.   - 5.
```

2.1.10 Answer the following
   1. Generate 10 equally spaced points between -1 and 1 in a column format.
   2. Generate a vector of points between 0 and 18 with step size 3.
   3. Generate a vector of points between 6 and -9 with step size -3.

2.1.11 List odd and even numbers less than 30 and put them in a two columns matrix and check the size.

```
--> x=1:2:30; y=2:2:30;
--> B=[x' y']
 B  =
      1.     2.
      3.     4.
      .........
     27.    28.
     29.    30.
--> size(B)
  ans  =
     15.    2.
```

2.1.12 Use Scilab to generate
  1. Odd numbers between 80 and 100.
  2. The numbers $30, 45, 60, \cdots, 150$

2.1.13 Use Scilab to answer:
  1. What are the first and last output generated by $\mathsf{linspace}(0,\pi)$ and how many outputs in all?
  2. Use the $\mathsf{linspace}$ command to generate the vector V: $0.2, 0.3, 0.4, 0.5, 0.6, 0.7, 0.8$

2.1.5 Empty Vectors

An **empty** vector is a vector that stores no values, can be created using empty square brackets:

```
--> vec = [ ]
  vec  =
      []
--> length(vec)
  ans  =
     0.
--> size(vec)
  ans  =
     0.    0.
```

We will use empty vectors to delete parts of a matrix in the next section.

2.1.6 Addressing and Subsetting Matrices

The address of an element in a vector v is its position in the row (or column) that is $v(k)$. The first position is 1. For example,

```
-> v = 2:3:15
  v  =
     2.    5.    8.    11.    14.
--> v(2)
  ans  =
     5.
--> v($)
  ans  =
```

14.

Note that $ is used to get the end of an array.

The address of an element in a matrix is its position A, defined by the row number and the column number where it is located $A(i, j)$. To address (extract) a part of an array, we use the : (colon) operator. The following table points to few options:

A(i , j)   returns the element a_{ij}
A(: , j)   returns the jth column of A
A(i , :)   returns the ith row of A
A(: , j : k)   returns a submatrix of the columns j,j+1, ..., k of A
A(:)   returns A as one column by concatenating the columns of A
A(: , [c1, c2, c3])   returns a submatrix of the columns c1, c2, c3 of A

2.1.14 Define the matrix $A = \begin{bmatrix} 1 & 2 & 3 & 4 & 5 \\ 2 & 3 & 4 & 5 & 6 \\ 3 & 4 & 5 & 6 & 7 \\ 4 & 5 & 6 & 7 & 8 \end{bmatrix}$

1. Create A using: operator and row by row.
2. Extract 5 from the second row.
3. Extract the second column of A.
4. Extract the second and fourth rows of A.
5. Extract the upper right 2 by 2 matrix
6. Extract the matrix of all interior elements.
7. Extract the last element in A.

Solution:

```
1. --> A = [1:5; 2:6; 3:7; 4:8]
   A  =
       1.    2.    3.    4.    5.
       2.    3.    4.    5.    6.
       3.    4.    5.    6.    7.
       4.    5.    6.    7.    8.
2. --> A(2,4)
   ans  =
       5.
3. A(:, 2)
4. --> A([2, 4], :)
   ans  =
       2.    3.    4.    5.    6.
       4.    5.    6.    7.    8.

5. --> A(1:2,4:5)
   ans  =
       4.    5.
       5.    6.
```

6. `--> A(2:3,2:4)`
```
   ans  =
      3.    4.    5.
      4.    5.    6.
```

7. `--> A($)`
```
   ans  =
      8.
```

☐

2.1.15 Enter the matrix $A = \begin{bmatrix} -2 & 1 & 4 & 1 \\ 4 & 0 & 2 & -5 \\ 17 & 3 & -1 & 8 \end{bmatrix}$

1. Create the matrix A
2. Replace -5 to 25
3. Add the fourth row of ones to A
4. Extract the first row
5. Remove second row and last column from A

Solution:

1. `--> A = [-2, 1, 4, 1; 4, 0, 2, -5; 17, 3, -1, 8]`
```
   A  =
     - 2.      1.      4.      1.
       4.      0.      2.    - 5.
      17.      3.    - 1.      8.
```
2. `A(2, 4) = 25`
3. `--> A = [A; 1, 1, 1,1]`
```
   A  =
     - 2.      1.      4.      1.
       4.      0.      2.     25.
      17.      3.    - 1.      8.
       1.      1.      1.      1.
```

4. `A(1, :)`
5. `--> A(:, [2,4]) = []`
```
   A  =
     - 2.      4.
       4.      2.
      17.    - 1.
       1.      1.
```

☐

2.1.7 Building New Matrices

The **ones(m,n)**, **zeros(m,n)**, **eye(m,n)**, **diag(vec)**, **matrix(v,m,n)** commands can be used to create matrices that have elements with special values .

ones (m,n)	returns an $m \times n$ matrix of ones
zeros (m,n)	returns an $m \times n$ matrix of zeros
zeros (A)	returns a matrix of zeros of size A
eye (m,n)	returns an $m \times n$ identity matrix
eye (A)	returns identity matrix of size A
diag(vec)	returns a square matrix with vec as its main diagonal
matrix(v, m,n)	returns an $m \times n$ matrix, by reshaping the vector v columnwise

2.1.16 Try out the construction of matrices using Scilab built-in commands:

```
1. --> M = ones(3,5)
   M   =
       1.    1.    1.    1.    1.
       1.    1.    1.    1.    1.
       1.    1.    1.    1.    1.


2. --> M = eye(3,3)
   M   =
       1.    0.    0.
       0.    1.    0.
       0.    0.    1.


3. --> M2 = zeros(2,6)
   M2  =
       0.    0.    0.    0.    0.    0.
       0.    0.    0.    0.    0.    0.


4. --> M3 = ones(M2)
   M3  =
       1.    1.    1.    1.    1.    1.
       1.    1.    1.    1.    1.    1.


5. --> vec =1:3
   vec  =

       1.    2.    3.
   --> M = diag(vec)
   M   =
       1.    0.    0.
       0.    2.    0.
       0.    0.    3.


6. --> x = 1:20;
   --> A = matrix(x, 4,5)
```

```
A   =
    1.    5.    9.    13.    17.
    2.    6.    10.   14.    18.
    3.    7.    11.   15.    19.
    4.    8.    12.   16.    20.
```

2.1.17 Construct the following matrices using special Scilab built-in commands:

1. $A = \begin{bmatrix} 4 & 4 & 4 & 4 \\ 4 & 4 & 4 & 4 \end{bmatrix}$

2. $A = \begin{bmatrix} 0 & 5 & 5 & 5 \\ 5 & 0 & 5 & 5 \\ 5 & 5 & 0 & 5 \\ 5 & 5 & 5 & 0 \end{bmatrix}$

3. $A = \begin{bmatrix} 2.1 & 1.5 & 1 & 1 \\ 3.4 & 0.6 & 1 & 1 \\ 1 & 0 & 0 & 0 \\ 0 & 1 & 0 & 0 \end{bmatrix}$

Solution:

```
1. A = 4 * ones(2,4)
2. A = 5 * (ones(4,4)- eye(4,4))
3. --> m1 = [2.1, 1.5; 3.4, 06]; m2 = ones(2,2);
   --> m3 = eye(m1); m4 = zeros(m1);
   --> A = [m1, m2; m3, m4]
    A   =
        2.1    1.5    1.    1.
        3.4    6.     1.    1.
        1.     0.     0.    0.
        0.     1.     0.    0.
```

□

2.1.18 Construct the matrix using Scilab built-in commands:

$$A = \begin{bmatrix} 1 & 2 & 3 & \cdots & 20 \\ 2 & 2 & 0 & \cdots & 0 \\ 3 & 0 & 3 & \cdots & 0 \\ \vdots & \vdots & \vdots & \ddots & \vdots \\ 20 & 0 & 0 & 0 & 20 \end{bmatrix}$$

Solution:

```
--> x=1:20;
--> A = diag(x); A(1,:)=x; A(:,1)=x'; A
```

□

2.2 Matrix Operations

Most of Scilab operations can be applied to matrices. These are standard arithmetic operations that is **addition, subtraction, multiplication, division, power, and transposition**:

$$+, \quad -, \quad *, \quad \wedge, \quad /, \quad \backslash, \quad '$$

2.2.1 For the matrices:

$$A = \begin{bmatrix} 1 & 2 \\ 3 & 4 \end{bmatrix}, B = \begin{bmatrix} 5 & 6 \\ 7 & 8 \end{bmatrix}.$$

Perform the following operations:

1. $A + B, A - B$
2. $A * B, B * A$ are they equal?
3. $A / B, A \backslash B$
4. $A * A * A, A^3$
5. $B', (B')'$
6. $A + 10, 10 * A$

Solution:

```
1. --> A = [1, 2; 3, 4], B = [5, 6; 7, 8]
   A   =

        1.    2.
        3.    4.
   B   =

        5.    6.
        7.    8.
  --> A + B,  A - B
   ans   =

        6.    8.
       10.   12.
   ans   =
      - 4.   - 4.
      - 4.   - 4.


2. --> A * B,  B * A
   ans   =

       19.    22.
       43.    50.
   ans   =

       23.    34.
       31.    46.


3. --> A/B , A\B
   ans   =

        3.   - 2.
        2.   - 1.
```

```
        ans   =
         - 3.   - 4.
           4.     5.

  4. --> A * A * A , A^3
        ans   =
           37.     54.
           81.    118.
        ans   =
           37.     54.
           81.    118.

  5. --> B, B' , (B')'
         B   =
            5.     6.
            7.     8.
        ans   =
            5.     7.
            6.     8.
        ans   =
            5.     6.
            7.     8.

  6. --> A + 10 , 10 * A
        ans   =
           11.     12.
           13.     14.
        ans   =
           10.     20.
           30.     40.
```

□

2.2.2 For the vectors $U = [1, 2, 3, 4]$ and $V = [1, 10, 100, 1000]$ and the matrices

$$A = \begin{bmatrix} 1 & 2 \\ 3 & 4 \\ 5 & 6 \end{bmatrix} \text{ and } B = \begin{bmatrix} 7 & -8 & 9 \\ -10 & 11 & -12 \end{bmatrix}$$

1. Check the size of each factor to predict the size of the product if it is defined.
2. $U * V, U * V'$
3. $U' * V,$
4. $A * B, A * B'$
5. $B * A,$

   Solution:

```
  1. --> U = 1:4 , V = [1, 10, 100, 1000]
         U   =
```

```
              1.      2.       3.      4.
      V   =
              1.      10.     100.    1000.
  --> [size(U) size(V)]
   ans  =
              1.      4.      1.      4.
  --> U * V
          !--error 10
  Inconsistent multiplication.
  --> [size(U) size(V')]
   ans  =
              1.      4.      4.      1.
  --> U * V'
   ans  =
          4321.


2. --> [size(U') size(V)]
   ans  =
          4.      1.      1.      4.
  --> U' * V
   ans  =
          1.      10.     100.    1000.
          2.      20.     200.    2000.
          3.      30.     300.    3000.
          4.      40.     400.    4000.


3. --> A = [1,2;3,4;5,6], B = [7, -8, 9; -10, 11, -12]
   A   =
          1.      2.
          3.      4.
          5.      6.
   B   =
          7.    - 8.       9.
       - 10.     11.    - 12.
  --> [size(A) size(B)]
   ans  =
          3.      2.      2.      3.
  --> A * B
   ans  =
        - 13.     14.    - 15.
        - 19.     20.    - 21.
        - 25.     26.    - 27.
  --> [size(A) size(B')]
   ans  =
          3.      2.      3.      2.
```

```
4. --> [size(B) size(A)]
    ans  =
      2.    3.    3.    2.
  --> B * A
    ans  =
      28.    36.
    - 37.  - 48.
```

□

Matrix Powers

Raising a square matrix A to power n is accomplished by successive multiplication. For example, $A^3 = AAA$. In Scilab, we use the power command: `A^2, A^3, A^4, ... .`

2.2.3 For the two special matrices $A = \begin{bmatrix} 1 & 1 \\ 0 & 1 \end{bmatrix}, B = \begin{bmatrix} 2 & -1 \\ 3 & -2 \end{bmatrix}$

Perform numerical explorations to establish the resulting pattern.

1. Compute A^2, A^3, A^{10} and A^{100}. Make a generalization about A^n for any natural number n.
2. Compute B^2, B^3, B^4 and B^5. Make a hypothesis about B^n as n increases.

2.2.4 For the matrices

$$A = \begin{bmatrix} 1 & 2 \\ 3 & 4 \\ 5 & 6 \end{bmatrix} \text{ and } B = \begin{bmatrix} 4 & 2 & 1 \\ 1 & 4 & 1 \\ 1 & 2 & 4 \end{bmatrix}$$

Find 2^A, 3^B, $\exp(A)$ and $\sqrt{B}$.

Solution:

```
--> A = [1,2;3,4;5,6], B = [4, 2, 1; 1, 4, 1; 1, 2, 4]
 A  =
      1.    2.
      3.    4.
      5.    6.
 B  =
      4.    2.    1.
      1.    4.    1.
      1.    2.    4.
  --> 2^A
    ans  =
      2.    4.
      8.    16.
      32.    64.
  --> 3^B
    ans  =
      81.    9.    3.
      3.    81.    3.
      3.    9.    81.
```

```
--> exp(A)
  ans  =
       2.718     7.389
       20.09     54.6
       148.4     403.4
--> sqrt(B)
  ans  =
       2.      1.414    1.
       1.      2.       1.
       1.      1.414    2.
```

Note the elementwise operations in this example.   □

Built-in Matrix Functions

Scilab has three main power functions defined for square matrices:
  1. The function *expm(A)* Defined as the expansion

$$e^A = I + A + \frac{A^2}{2!} + \frac{A^3}{3!} + \cdots$$

where the Pade approximation is used.
  2. The function *sqrtm(A)* computes $A^{1/2}$. The square root is unique for symmetric and positive matrices.
  3. The function *logm(A)* computes the natural logarithm of A.

2.2.5 For the matrix
$$A = \begin{bmatrix} 4 & 1 \\ 1 & 9 \end{bmatrix}$$
Compare $\exp(A)$ and *expm(A)*, $\sqrt{A}$ and *sqrtm(A)*.

  Solution:

```
--> A = [4, 1;1,9]
  A  =
       4.       1.
       1.       9.
--> exp(A) , expm(A)
  ans  =
       54.6      2.718
       2.718     8103.
  ans  =
       394.7     1816.
       1816.     9474.
--> sqrt(A), sqrtm(A)
  ans  =
       2.       1.
```

```
     1.    3.
  ans  =
     1.99      0.201
     0.201     2.993
```

□

2.2.1 Constructing New Matrices

A **random matrix** is a matrix whose elements are random numbers. In Scilab, **random matrices** can be generated by the rand function which produces uniformly distributed random number between 0 and 1. In addition to the **default uniform** random numbers, the rand function has an option rand('normal') to create **normally distributed** random numbers with a mean 0 and a variance 1.

The following table summarizes the rand function and some of its options:

rand	returns uniformly distributed random number between 0 and 1
rand(1, n)	returns a row vector with uniformly distributed random numbers between 0 and 1
rand(v)	returns a row vector with uniformly distributed random numbers between 0 and 1 with length equals to the vector v
rand(n,m)	returns an $n \times m$ matrix with entries uniformly distributed random numbers between 0 and 1
rand(n,m, 'uniform')	returns an $n \times m$ matrix with entries uniformly distributed random numbers between 0 and 1
rand('normal')	returns normally distributed random number with a mean 0 and a variance 1
randn(n,m,'normal')	returns an $n \times m$ matrix with entries normally distributed random numbers

In the following problem, we will explore matrix construction using the rand function.

2.2.6 Answer the following:
1. Generate 10 uniformly distributed random numbers.
2. Generate a 10 by 10 matrix of integers uniformly distributed between 0 and 10.
3. Generate a 5 by 5 matrix of normally distributed entries.
4. Use the commands t=rand(2250,1); histplot(15,t). What is the shape of the histogram? What happens if you run t=rand(2250000,1); histplot(15,t)
5. s=rand(10000,1,'normal'); histplot(20,s). What is the shape of the histogram?

Solution:

```
1. -->format(6)
   --> rand(1, 10)
     ans  =
         0.160     0.666     0.433     0.967     0.190     0.756     0.237
         0.702     0.053     0.514
```

*These random numbers are pseudo-random (generated by an algorithm). The function rand has an option called **seed**, we can specify the seed in order to get the same numbers of random numbers repeatedly.*

```
-> rand("seed", 135) // You can pick any number as a seed
--> rand(1, 10)
  ans  =
      0.226     0.765     0.664     0.470     0.678     0.669     0.123
      0.595     0.056     0.695
--> rand("seed", 135)
--> rand(1, 10)
  ans  =
      0.226     0.765     0.664     0.470     0.678     0.669     0.123
      0.595     0.056     0.695
```

2. *We show the first two rows:*
```
--> rand("seed", 2020)
--> round( 10 * rand(10, 10))
  ans  =
      5.    9.    5.    8.    6.    9.    8.    10.    6.    2.
      6.    3.    3.    9.    6.    0.    8.    3.     6.    8.
```

3.
```
--> rand("seed", 123)
--> rand( 5, 5, 'normal')
   ans  =
       0.093  - 0.937     0.800  - 0.71      1.918
       1.698  - 1.53      0.580    0.061     0.341
       1.421    0.703     1.249  - 0.372     0.129
       0.990    1.126     0.195    1.387     1.988
     - 0.814    0.087     0.672  - 0.284   - 2.099
```

4.
```
--> t = rand(2250, 1); histplot(15, t);
--> t = rand(2250000, 1); histplot(15, t);
```

5.
```
--> t = rand(10000, 1, 'normal'); histplot(20, t);
```

□

2.2.7 Construct the following matrices:

1. Generate a 3 by 5 random matrix of uniformly distributed numbers between 0 and 10.

2. Generate a 4 by 4 random matrix of uniformly distributed numbers between 0 and 9.

Solution:

```
1. --> A - round(10*rand(3,5))
   A  =
        2.    6.    1.    1.    6.
        0.    10.   7.    6.    3.
        3.    6.    6.    7.    2.

2. --> A = int(10*rand(4,4))
   A  =
        1.    9.    6.    5.
        0.    5.    0.    3.
        3.    1.    6.    1.
        5.    6.    7.    6.
```

□

2.2.2 Elementwise Arithmetic Operations

The elementwise operations are defined on matrices of equal sizes. The addition and subtraction were elements by element defined. However, we have three new operations on matrices:

- .∗ : dot(or period) product
- ./ : dot(or period) division
- .∧ : dot(or period) power

Let us explore these operations:

2.2.8 Define the two matrices:

```
--> A = [1 2 3; 4 5 6], B = [ 10, 5, -1; 3, -2, 1]
 A  =
     1.    2.    3.
     4.    5.    6.
 B  =
     10.    5.   - 1.
     3.   - 2.    1.
```

Compute
1. $A.*B$
2. $A./B$
3. $A.\backslash B$
4. $A.\wedge B$
5. $A.*A, A.\wedge 2, A*A$
6. $B.\wedge 3, 1./B$

Solution:

```
1. -->   A .* B
   ans  =
        10.    10.   - 3.
        12.   - 10.    6.
```

```
2. --> A ./ B
   ans  =
       0.1       0.4   - 3.
       1.333   - 2.5     6.
3. --> A .\ B
   ans  =
       10.       2.5   - 0.333
       0.75    - 0.4     0.167
4. --> A .^ B
   ans  =
       1.       32.      0.333
       64.      0.04     6.
5. --> A .* A
   ans  =
       1.       4.       9.
       16.      25.      36.
   --> A .^2
   ans  =
       1.       4.       9.
       16.      25.      36.
   --> A * A
           !--error 10
   Inconsistent multiplication.
6. --> B .^3
   ans  =
       1000.     125.   - 1.
       27.     - 8.       1.
   --> 1 ./ B
   ans  =
       0.1       0.2   - 1.
       0.333   - 0.5     1.
```
□

2.2.9 Define the vectors $x = \begin{bmatrix} 1 & 2 & 3 & 4 & 5 & 6 & 7 \end{bmatrix}$ and $y = \begin{bmatrix} -3 & -2 & -1 & 0 & 1 & 2 & 3 \end{bmatrix}$.

The vectors x and y can be thought as a grid of points on x-axis and y-axis. Perform and explain the results of the following operations:

```
1. --> x = 1:7,  y = -3:3
   x  =
       1.    2.    3.    4.    5.    6.    7.
   y  =
     - 3.  - 2.  - 1.    0.    1.    2.    3.
2. --> f = x.^2
   f  =
       1.    4.    9.    16.    25.    36.    49.
   --> f = sin(x)./x
   f  =
       0.841    0.455    0.047  - 0.189  - 0.192  - 0.047    0.094
```

```
3. --> z = x + y
   z  =
     - 2.    0.    2.    4.    6.    8.    10.
   --> z = x.^2 + 2*x.*y + y.^2
   z  =
       4.    0.    4.    16.    36.    64.    100.
4. --> region = x' * y
   region  =
     - 3.   - 2.   - 1.    0.    1.    2.    3.
     - 6.   - 4.   - 2.    0.    2.    4.    6.
     - 9.   - 6.   - 3.    0.    3.    6.    9.
     - 12.  - 8.   - 4.    0.    4.    8.    12.
     - 15.  - 10.  - 5.    0.    5.    10.   15.
     - 18.  - 12.  - 6.    0.    6.    12.   18.
     - 21.  - 14.  - 7.    0.    7.    14.   21.

   --> x*y'
   ans  =
       28.
   --> x.*y
   ans  =
     - 3.   - 4.   - 3.    0.    5.    12.    21.
   --> y'.*x'
   ans  =

     - 3.
     - 4.
     - 3.
       0.
       5.
      12.
      21.
5. --> y ./ x
   ans  =
     - 3.   - 1.   - 0.333    0.    0.2    0.333    0.429
```

Comparison Operators

Matrices can be compared using the logical operators: $<, \leq, >, \geq, ==, \sim=$. The results of comparisons are T or F.

2.2.10 Build two 4 by 4 matrices of uniformly random integers between 0 and 10, the fist matrix A with seed 135 and the second matrix B with seed 246. Perform the comparisons $A > B$, $A == B$, $A \sim= 0$.

Solution:

```
--> rand("seed", 135), A = round( 10 * rand(4,4))
 A   =
      2.     7.     1.     9.
      8.     7.     7.     1.
      7.     1.     7.     0.
      5.     6.     7.     7.
--> rand("seed", 246), B = round( 10 * rand(4,4))
 B   =
      8.     9.     8.     9.
      7.     9.    10.     9.
      4.     7.     1.    10.
      9.     5.     8.     2.
--> A > B
 ans   =
   F F F F
   T F F F
   T F T F
   F T F T
--> A == B
 ans   =
   F F F T
   F F F F
   F F F F
   F F F F
--> A ~= 0
 ans   =
   T T T T
   T T T T
   T T T F
   T T T T
```

□

2.2.3 Vector Products

Given the two vectors $a = [a_1, a_2, a_3]$ and $b = [b_1, b_2, b_3]$.

The dot product is defined by the scalar $a \cdot b = a_1 b_1 + a_2 b_2 + a_3 b_3$ and

the cross product is the vector given by $a \times b = (a_2 b_3 - a_3 b_2 \quad, a_3 b_1 - a_1 b_3 \quad, a_1 b_2 - a_2 b_1)$

The **dot (scalar or inner) product** of any two vectors U and V of the same dimensions is accomplished with the command $sum(U. * V)$ and the **cross (vector) product** of any two vectors of dimensions 3 by 1 is accomplished with the command $cross(U, V)$

2.2.11 Compute the dot product and the cross product of the vectors $U = [2, \quad 4, \quad 6]$ and $V = [3, \quad -5, \quad 7]$

Solution:

```
--> U = [2, 4, 6], V = [3, -5, 7]
 U  =
    2.    4.    6.
 V  =
    3.  - 5.    7.
-->dotUV = sum(U.*V)
 dotUV  =
    28.
-->crossUV = cross(U,V)
 crossUV  =
    58.    4.  - 22.
```

□

2.2.12 Find the angle between the vector $a = [3, \quad 2, \quad 1]$ and $b = [6, \quad -3, \quad -3]$. Use the formula

$$\cos(\theta) = \frac{a \cdot b}{\sqrt{|a|}\sqrt{|b|}}$$

where $a \cdot b$ is the dot product and $|a|, |b|$ are the lengths of vectors a and b.

Solution:

```
--> a = [3 2 1]; b = [6 -3 -3];
--> dot_ab = sum(a.*b)
 dot_ab  =
    9.
-->len_a = sqrt (sum(a.*a))
 len_a  =
    3.742
-->len_b = sqrt (sum(b.*b))
 len_b  =
    7.348
--> cos_ab = dot_ab/(len_a * len_b)
 cos_ab  =
    0.327
--> theta_r = acos(cos_ab) \\ the angle in radians
 theta_r  =
    1.237
--> theta_d = acosd(cos_ab) \\ the angle in degrees
 theta_d  =
    70.89
```

□

2.2.4 New Matrices from Given Matrices

Scilab has commands to create new matrices using parts of already constructed matrices. For a given matrix $A^{m \times n}$, and X is a vector with n components, the commands diag, triu, tril create new matrices as in the table

diag(A)   returns a column vector of the main diagonal of A

diag(X)   returns a square matrix $n \times n$ with X as its main diagonal and zero elsewhere.

diag(A,k)   returns a column vector of the k diagonal of A where $k = 0$ is the main diagonal and $k > 0$ from over the main and $k < 0$ from below the main.

diag(X,k)   returns $(n + abs(k)) \times (n + abs(k))$ matrix with vector X on the kth diagonal.

triu(A)   returns an upper triangular matrix of A size.

tril(A)   returns a lower triangular matrix of A size.

2.2.13 Suppose $A = \begin{bmatrix} 1 & 5 & 9 & 13 \\ 2 & 6 & 10 & 14 \\ 3 & 7 & 11 & 15 \\ 4 & 8 & 12 & 16 \end{bmatrix}$ and the vector $X = [10, \quad 20, \quad 30, \quad 40]$

Explain the output of the following commands on A and X:

1. *diag(A), diag(diag(A))*
2. *diag(A,1), diag(diag(A,1),1)*
3. *diag(A,-2), diag(A,-2),-2)*
4. *diag(X), diag(X')*
5. *triu(A), tril(A)*

Solution: *We define A and X:*

```
--> A = matrix(1:16, 4, 4), X = 10:10:40
 A   =
    1.      5.      9.      13.
    2.      6.      10.     14.
    3.      7.      11.     15.
    4.      8.      12.     16.
 X   =
    10.     20.     30.     40.

1. --> diag(A), diag(diag(A))
    ans   =
       1.
       6.
       11.
       16.
    ans   =
       1.      0.      0.      0.
       0.      6.      0.      0.
       0.      0.      11.     0.
       0.      0.      0.      16.
```
main diagonal as a vector versus a matrix format.
```
2. --> diag(A,1), diag(diag(A,1),1)
    ans   =

       5.
```

```
      10.
      15.
   ans   =
      0.      5.      0.      0.
      0.      0.     10.      0.
      0.      0.      0.     15.
      0.      0.      0.      0.
```
An upper main diagonal in vector and matrix format.

3. --> diag(A,-2), diag(diag(A,-2),-2)
```
   ans   =
      3.
      8.
   ans   =
      0.      0.      0.      0.
      0.      0.      0.      0.
      3.      0.      0.      0.
      0.      8.      0.      0.
```
A lower diagonal.

4. --> diag(X), diag(X')
```
   ans   =
     10.      0.      0.      0.
      0.     20.      0.      0.
      0.      0.     30.      0.
      0.      0.      0.     40.
   ans   =
     10.      0.      0.      0.
      0.     20.      0.      0.
      0.      0.     30.      0.
      0.      0.      0.     40.
```

5. --> triu(A), tril(A)
```
   ans   =
      1.      5.      9.     13.
      0.      6.     10.     14.
      0.      0.     11.     15.
      0.      0.      0.     16.
   ans   =
      1.      0.      0.      0.
      2.      6.      0.      0.
      3.      7.     11.      0.
      4.      8.     12.     16.
```

□

2.2.14 Construct the matrix without typing the entries: $A = \begin{bmatrix} 1 & 4 & 0 & 0 & 0 \\ -2 & 2 & 3 & 0 & 0 \\ 0 & -3 & 3 & 2 & 0 \\ 0 & 0 & -4 & 4 & 1 \\ 0 & 0 & 0 & -5 & 5 \end{bmatrix}$

Solution: *We construct the diagonals:*

```
--> d = 1:5, du = 4:-1:1, dl = -(2:5)
 d   =
    1.    2.    3.    4.    5.
 du  =
    4.    3.    2.    1.
 dl  =
  - 2.  - 3.  - 4.  - 5.
--> A = diag(x) + diag(xu,1) + diag(xl, -1)
 A   =
    1.    4.    0.    0.    0.
  - 2.    2.    3.    0.    0.
    0.  - 3.    3.    2.    0.
    0.    0.  - 4.    4.    1.
    0.    0.    0.  - 5.    5.
```

$\square$

2.3 Systems of Linear Equations

Scilab is an ideal environment for exploring the linear system of equations because Scilab functions and commands are designed to work directly on vectors and matrices. Scilab is rich in functions and operators which can simplify and manipulate matrices. Linear systems of equations are the most common numerical problems. Usually, Scilab solves systems of linear equations by left division operator $\backslash$.

2.3.1 Basic Functions of Matrices

The following functions are used to compute the determinant, the rank, the inverse, and the trace of a matrix A.

det(A)	returns the determinant of a square matrix A
rank(A)	returns the rank of A that is the number of linearly independent columns and rows
inv(A)	returns the inverse of a square matrix A
pinv(A)	returns the pseudo-inverse of a matrix A. The pseudo-inverse of an $n \times m$ matrix is an $m \times n$
trace (A)	returns the determinant of a square matrix A

2.3.1 Consider the three matrices: $A = \begin{bmatrix} 1 & 2 \\ 3 & 4 \end{bmatrix}$, $B = \begin{bmatrix} 1 & 2 \\ 5 & 10 \end{bmatrix}$, and $C = \begin{bmatrix} 1 & 2 & 3 \\ 2 & 3 & 5 \end{bmatrix}$.

Apply the commands and compare:
1. *det(A) and det(A'), det(B), det(C)*
2. *inv(A) and inv(A'), inv(B), inv(C)*
3. *pinv(A), pinv(B), pinv(C)*
4. *rank(A) and rank(A'), rank(B) and rank(B'), rank(C) and rank(C')*
5. *trace(A), trace(B), trace(C)*

Solution: *We define the matrices:*

```
--> A = [1, 2; 3, 4], B = [1, 2; 5, 10], C = [1,2,3; 2, 3, 5]
 A   =
    1.    2.
    3.    4.
 B   =
    1.    2.
    5.    10.
 C   =
    1.    2.    3.
    2.    3.    5.
```

1.
```
 --> det(A), det(A')
    ans   =
     - 2.
    ans   =
     - 2.
 --> det(B)
    ans   =
```

```
          0.
   -->det(C) // Note the determinant is only defined for square matrices.
          !--error 20
   Wrong type for first argument: Square matrix expected.
2. --> inv(A), inv(A')
     ans  =
      - 2.      1.
        1.5  - 0.5
     ans  =
      - 2.      1.5
        1.   - 0.5
   --> inv(B)
          !--error 19
   Problem is singular.
   -->inv(C)
          !--error 20
   Wrong type for first argument: Square matrix expected.
3. --> pinv(A)
     ans  =
      - 2.      1.
        1.5  - 0.5
   -->pinv(B)
     ans  =
        0.008    0.038
        0.015    0.077
   -->pinv(C)
     ans  =
      - 2.667    1.667
        2.333  - 1.333
      - 0.333    0.333
4. --> rank(A), rank(A')
     ans  =
        2.
     ans  =
        2.
   --> rank(B), rank(B')
     ans  =
        1.
     ans  =
        1.
   --> rank(C), rank(C')
     ans  =
        2.
     ans  =
        2.
5. --> trace(A), trace(B), trace(C)
```

```
        ans   =
           5.
        ans   =
            11.
        !--error 10000
       trace: Wrong size for input argument #1: A square matrix expected.
       at line       7 of function trace called by :
        trace(A), trace(B), trace(C)
```

□

2.3.2 For the matrix $A = \begin{bmatrix} 0 & 1 & 1 & 1 & 5 \\ 1 & 0 & 1 & 1 & 1 \\ 1 & 1 & 0 & 1 & 1 \\ 1 & 1 & 1 & 0 & 1 \\ 5 & 1 & 1 & 1 & 0 \end{bmatrix}$,

Compute
  1. d=det(A) , r=rank(A) , and t=trace (A)
  2. B=inv(A) . Check the matrix.

Solution: *We define the matrix A as follows:*

```
--> A = ones(5,5)-eye(5,5)
 A   =
    0.    1.    1.    1.    1.
    1.    0.    1.    1.    1.
    1.    1.    0.    1.    1.
    1.    1.    1.    0.    1.
    1.    1.    1.    1.    0.
--> A(1,5) =5; A(5,1) = 5
 A   =
    0.    1.    1.    1.    5.
    1.    0.    1.    1.    1.
    1.    1.    0.    1.    1.
    1.    1.    1.    0.    1.
    5.    1.    1.    1.    0.

1. --> d = det(A), r = rank(A), t = trace(A)
    d   =
     - 20.
    r   =
       5.
    t   =
       0.
2. --> B = inv(A), B*A
    B   =
       0.15   - 0.25   - 0.25   - 0.25    0.35
```

```
       -  0.25   -  0.25      0.75      0.75   -  0.25
       -  0.25      0.75   -  0.25      0.75   -  0.25
       -  0.25      0.75      0.75   -  0.25   -  0.25
          0.35   -  0.25   -  0.25   -  0.25      0.15
```
The product of BA results in an identity matrix.

□

2.3.2 Solving Linear Equations

The general form of linear system of equations can be written in matrix format as

$$Ax = b$$

where A is the coefficient matrix and b is the right hand side and x is the vector of unknown variables, or in componentwise, as

$$
\begin{aligned}
a_{11}x_1 + a_{12}x_2 + a_{13}x_3 + \cdots + a_{1m}x_m &= b_1 \\
a_{21}x_1 + a_{22}x_2 + a_{23}x_3 + \cdots + a_{2m}x_m &= b_2 \\
a_{31}x_1 + a_{32}x_2 + a_{33}x_3 + \cdots + a_{3m}x_m &= b_3 \\
\cdots &= \cdots \\
a_{n1}x_1 + a_{n2}x_2 + a_{n3}x_3 + \cdots + a_{nm}x_m &= b_n
\end{aligned}
$$

To solve the linear system of equations $Ax = b$, where A is the coefficient matrix, b is a given vector, the unknown vector solution x is equal to $x = A^{-1}b$ if the inverse matrix A^{-1} exists. We also can compute the solution using x=A\b

2.3.3 Solve the matrix equation $A = \begin{bmatrix} 2 & 2 & -1 & 1 \\ 4 & 3 & -1 & 2 \\ 8 & 5 & -3 & 4 \\ 3 & 3 & -2 & 2 \end{bmatrix} \begin{bmatrix} x \\ y \\ z \\ w \end{bmatrix} = \begin{bmatrix} 4 \\ 6 \\ 12 \\ 6 \end{bmatrix}$

Solution:

```
--> A=[2 2 -1 1; 4 3 -1 2; 8 5 -3 4; 3 3 -2 2]
A =

       2       2      -1       1
       4       3      -1       2
       8       5      -3       4
       3       3      -2       2
--> b=[4;6;12;6];
--> X=inv(A)*b
X =

       1
```

```
          1
         -1
         -1
--> A*X // Check the solution AX=b
 ans =

          4
          6
         12
          6
--> X=A\b // Using the division
 X =

          1.0000
          1.0000
         -1.0000
         -1.0000
```

□

Gauss-Jordan Elimination

For the general matrix equation $Ax = b$, where A can be a rectangular matrix, we combine the right side with the matrix A to get the **augmented matrix** and we calculate the reduced row echelon form to find the solution.

2.3.4 Solve the system

$$-2x + y + 2z = 4$$
$$x - 4y - 2z = -6$$
$$-x + 2y - 2z = 2$$

using Gauss-Jordan elimination.

Solution: *Write the augmented matrix and use rref function*

```
--> A=[-2 1 2 4; 1 -4 -2 -6; -1 2 -2 2], AR=rref(A)
 A =
        -2       1       2       4
         1      -4      -2      -6
        -1       2      -2       2
 AR =
        1.0000            0            0      -1.0000
             0       1.0000            0       1.0000
             0            0       1.0000      0.5000
```

The solution is the vector $(x, y, z) = (-1, 1, 0.5)$

□

2.3.5 Given the matrix $A = \begin{bmatrix} 2 & 2 & -1 & 1 \\ 4 & 3 & -1 & 2 \\ 8 & 5 & -3 & 4 \\ 3 & 3 & -2 & 2 \end{bmatrix}$, and the vector $b = [4, 6, 12, 6]'$.

Find the solution of $AX = b$.

Row Operations

Scilab has a default function to compute the reduced row echelon for (RREF). Scilab has a built-in function

`rref(A)`

to compute the RREF form of any matrix.

2.3.6 Determine the reduced row echelon form of A:

1. $A = \begin{bmatrix} 1 & 3 & 1 \\ -3 & -6 & 1 \\ 1 & 3 & 0 \end{bmatrix}$, $B = \begin{bmatrix} 1 & 2 & 3 \\ 4 & 6 & 0 \\ 2 & 3 & 0 \end{bmatrix}$, $C = \begin{bmatrix} 1 & 3 & 1 & 5 \\ -3 & -6 & 1 & 0 \\ 1 & 3 & 0 & 2 \end{bmatrix}$

2. $A = \begin{bmatrix} 3 & 1 & -2 & 14 \\ 3 & 1 & 1 & 15 \\ 2 & 0 & -2 & 11 \end{bmatrix}$

3. $A = \begin{bmatrix} -3 & 1 & -2 \\ 2 & 4 & -2 \\ 2 & 0 & 3 \\ 1 & 4 & 2 \end{bmatrix}$

Solution:

```
-->A=[1 3 1;-3 -6 1;1 3 0]; RA=rref(A)
 RA   =
    1.    0.    0.
    0.    1.    0.
    0.    0.    1.
-->B=[1 2 3; 4 6 0; 2 3 0]; RB=rref(B)
 RB   =
    1.    0.   - 9.
    0.    1.    6.
    0.    0.    0.
-->C=[1 3 1 5;-3 -6 1 0;1 3 0 2]; RC=rref(C)
 RC   =
    1.    0.    0.  - 1.
    0.    1.    0.    1.
    0.    0.    1.    3.
```

□

2.3.3 Eigenvalues and Eigenvectors

Eigenvalue problems arise in many branches of science and engineering applications. For an $n \times n$ square matrix A, the problem is to find the solution of the linear system of equations

$$Ax = \lambda x$$

where λ is a scalar and x is a non-zero column vector of length n. The scalar λ is called an **eigenvalue** of A, and x is called an **eigenvector**. in Scilab, the function *spec* computes the eigenvalues and the eigenvectors of matrix A.

$spec(A)$ returns the eigenvalues of A as a vector

$[V,D] = spec(A)$ returns A matrix V of eigenvectors and a matrix D of eigenvalues of A

2.3.7 Find the eigenvalues and eigenvectors of the matrix A:

$$A = \begin{bmatrix} 1 & -3 & 4 \\ 4 & -7 & 8 \\ 6 & -7 & 7 \end{bmatrix}$$

Solution:

```
--> A=[1 -3 4;4 -7 8; 6 -7 7]
A =
        1     -3      4
        4     -7      8
        6     -7      7
--> [V D]=spec(A)
V =
        0.3333    -0.4082    -0.4082
        0.6667    -0.8165    -0.8165
        0.6667    -0.4082    -0.4082
D =
        3.0000          0          0
             0    -1.0000          0
             0          0    -1.0000
```

$\square$

2.3.8 Find the eigenvalues and eigenvectors of the matrix A:

$$A = \begin{bmatrix} 3 & 4 & -2 \\ 3 & -1 & 1 \\ 2 & 0 & 5 \end{bmatrix}$$

Solution:

```
-->  A=[3 4 -2;3 -1 1; 2 0 5]
 A   =
        3.     4.    - 2.
        3.   - 1.      1.
        2.     0.      5.
```

```
--> [ V, D] = spec(A)
 D  =
   - 2.75        0                    0
     0           4.875 + 1.431i       0
     0           0                    4.875 - 1.431i
 V  =
     0.533      0.048 - 0.555i      0.048 + 0.555i
   - 0.835    - 0.167 - 0.243i    - 0.167 + 0.243i
   - 0.138    - 0.776             - 0.776
```

Scilab has a function poly to compute the characteristic polynomial

$$p(\lambda) = det(A - \lambda I) = \lambda^3 - 7\lambda^2 - \lambda + 71$$

We use s variable for the eigenvalue:

```
-->poly(A,"s")
 ans  =
                    2   3
     71 - s - 7s + s
```

□

Singular Value Decomposition

Scilab can compute the **singular value decomposition, SVD** and the singular values of matrices. The function *svd* computes the singular value decomposition components.

$svd(A)$	returns the singular values of A as a vector
$[U, S, V] = svd(A)$	returns A diagonal matrix S of the same size as A and two unitary matrices U, V such that $A = USV'$

2.3.9 Determine the singular values and the unitary matrices and check the answer for the matrix A:

$$A = \begin{bmatrix} 1 & 1 & 1 \\ 0 & 2 & 0 \\ -1 & 0 & -1 \\ 2 & 0 & -2 \end{bmatrix}$$

Solution:

```
--> A=[1, 1, 1; 0, 2, 0; -1, 0, -1; 2, 0, -2]
 A  =
       1.    1.    1.
       0.    2.    0.
     - 1.    0.  - 1.
       2.    0.  - 2.
--> s = svd(A)
```

```
 s   =
     2.828
     2.449
     1.732
--> [U, S, V] = svd(A)
 V   =
   - 0.707     0.408      0.577
   - 0.000     0.816    - 0.577
     0.707     0.408      0.577
 S   =
     2.828     0.         0.
     0.        2.449      0.
     0.        0.         1.732
     0.        0.         0.
 U   =
   - 0.000     0.667      0.333    - 0.667
   - 0.000     0.667    - 0.667      0.333
   - 0.000   - 0.333    - 0.667    - 0.667
   - 1.      - 0.000      0.000      0.000
--> U * S * V'
  ans   =


     1.         1.         1.
   - 0.000     2.       - 0.000
   - 1.      - 0.000    - 1.
     2.        0.000    - 2.
```

□

2.3.10 Solve the system

$$-2x+y+z = -1$$
$$3x+y+2z = 1$$
$$-5x+2x-z = 3$$

Verify that the obtained answer satisfies the system.

Solution:

```
-->A=[-2 1 1;3 1 2;-5 2 -1];b=[-1;1;3];
-->x=A\b
 x   =
     0.7857
     2.5
   - 1.9286
-->A*x
  ans   =
   - 1.
```

1.
3.

□

2.4 Homework: Matrices

Use Scilab to answer the following:

2.4.1   1. Generate 20 equally spaced points between 0 and π in a vector format.
2. Generate a vector of points between -12 and 12 with step size 2 and using : operator.

2.4.2 Find the length of the vector $V = (2 \quad -3 \quad 4 \quad -5)$.

2.4.3 Find the length of the vector $V = (-2 \quad 1+i \quad -3i \quad 3)$.

2.4.4 The components of a vector V are $-3, 2, 1-i, \pi$. Enter this vector in row and column format.

2.4.5 Create a 6×6 matrix A with twos on the diagonal and zeroes everywhere else.

2.4.6 Create a 7×7 matrix B with zeroes on the diagonal and threes everywhere else.

2.4.7 Compute the array product $A.*B$ and the matrix product AB of the matrices

$$A = \begin{bmatrix} 3 & -2 & 0 \\ 5 & -4 & 1 \\ 0 & 1 & -1 \end{bmatrix} \text{ and } B = \begin{bmatrix} 0 & 3 & 4 \\ 2 & -5 & 0 \\ -1 & 0 & 1 \end{bmatrix}$$

2.4.8 Find the eigenvalues and eigenvectors of the matrices A and B:

$$A = \begin{bmatrix} 3 & -1 & 0 & 0 \\ 1 & 1 & 0 & 0 \\ 3 & 0 & 5 & -3 \\ 4 & -1 & 3 & -1 \end{bmatrix} \text{ and } B = \begin{bmatrix} 1 & 2 & 3 \\ 4 & 5 & 6 \\ 7 & 8 & 9 \end{bmatrix}$$

2.4.9 Solve the linear system of equations

$$\begin{aligned} x + y - z &= 1 \\ 2x - y + z &= 2 \\ -4y + 3z &= 12 \end{aligned}$$

using three different methods
1. Matrix division method
2. Inverse matrix method
3. Augmented matrix (Reduced row echelon form) method

2.4.10 Solve the matrix equation $A = \begin{bmatrix} 1 & 4 & -3 & 4 \\ -1 & 1 & -1 & 4 \\ -2 & -1 & -1 & 4 \\ -2 & 0 & -1 & 2 \end{bmatrix} \begin{bmatrix} x \\ y \\ z \\ w \end{bmatrix} = \begin{bmatrix} 2 \\ 12 \\ -6 \\ -12 \end{bmatrix}$

2.4.11 Given the matrices

$$a = \begin{bmatrix} 10 & 14 & 0 & 1 \end{bmatrix}, b = \begin{bmatrix} 2 & 5 & 8 \\ 4 & 1 & -2 \\ 6 & 3 & 0 \end{bmatrix}, c = \begin{bmatrix} 20 \\ 10 \\ 1 \end{bmatrix}$$

1. Assign to $x1$ the second column of a
2. Assign to $x2$ the first column of b
3. Assign to $x3$ the third row of b
4. Assign to $x4$ the diagonal of b
5. Assign to $x5$ the second column of a and c

2.4.12 Using zeros ones diag functions, compute the following

1. Create 4×4 matrix of zeros
2. Create 2×4 matrix of zeros
3. Create 3×3 matrix of ones
4. Create 5×4 matrix of 6
5. Create 4×4 matrix of zeros
6. Create 4×4 matrix whose diagonal is $1, 2, 3, 4$

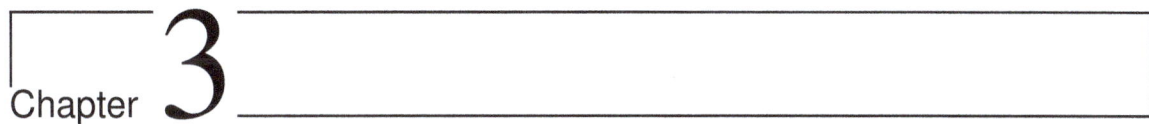

Chapter **3**

Programming

A **computer program** is a sequence of commands or instructions that accomplishes a task. To execute, or run, a program is to have the computer follow these instructions. **High-level languages** such as C, FORTRAN, and many more, have English-like commands and functions, such as "print n" or " for k= 1: 10 do this". The computer can interpret instructions only written in its machine language. Programs that are written in high-level languages must therefore be translated into machine language before the computer can execute the sequence of commands in the program. A program that does this translation from a high-level language to an executable file is called a **compiler**. The original program is called the source code, and the resulting executable program is called the **object code**.

On the other hand, an **interpreter** goes through the code line-by-line, executing each command as it goes. Scilab uses either what is called script files. These script files are **interpreted**, rather than compiled. Therefore, the correct terminology is that these are scripts and not programs. In this book, we will use the word *program* to mean a set of *scripts* and *functions*.

By conventions, the file extensions used in Scilab, are the suffix:

.sce for script files and **.sci** for functions files.

3.1 Script Files

- A script file is a sequence of Scilab instructions or commands.
- To run (or execute) a script file, Scilab will execute the commands in the order they are written.
- Any assigned output will be displayed in the console if there is no suppression (no semicolon at the end).
- Script files can be edited and improved.
- Script files can be documented with comments.
- Script files can be typed and edited in any text editor and Scilab has its dedicated editor.
- The extension for Scilab script files is .sce

Next, we will present several examples of scripts.

3.1.1 Find the surface area of a standard soccer ball with a diameter of 22 cm and find its volume. We will write a script file so solve this problem, we will call it *script1.sce*

```
// This program calculates the surface area and
// the volume of a soccer ball with diameter 22 cm
```

```
d = 22.0;         // The diameter
// The surface area and the volume formulas of a ball are
// S = 4 pi r^2
// V = 4/3 pi r^3
// r is the radius
r = d/2;                 // The radius
S = 4 * %pi * r^2;       // The surface area
V = 4/3 * %pi * r^3;     // The volume

// We use a nice formatted output
mprintf(' Diameter = %3.2f\n Surface = %3.2f\n Volume = %3.2f\n', d,S,V )
// or simply
[d, S, V]
```

Scripted files can be executed line by line or as a file. To run a file, select the file menu on the console, then execute the file, select script1.sce, and run.

3.1.2 **Math Puzzle 1**: Draw a circle inscribed in a square and a second square inscribed in the circle. Find the ratio of the area of the two squares. Here, we visualize the picture and the trick is to rotate the inner square by 90 degrees.

```
// circlepuzzle.sce
// Program to visualize the puzzle
//
// Plot a unit circle
t = 0:.01:2*%pi;
x = cos(t); y = sin(t);
square(-2,2,-2,2)
plot(x,y,'thickness',3)
//Plot an inscribed square
a=sqrt(2)/2;
x = a*[1, 1, -1,-1, 1];  //x--coordinates of vertices
y = a*[-1, 1, 1, -1, -1];//y--coordinates of vertices
plot(x,y,'r', 'thickness',4)
//Plot a circumscribed square
x = [1, 1, -1,-1, 1];//x--coordinates of vertices
y = [-1, 1, 1, -1, -1];//y--coordinates of vertices
plot(x,y,'k', 'thickness',4)
g=get("current_axes");//get the handle of the newly created axes
g.axes_visible="off"; // makes the axes visible
```

3.1.3 **Math Puzzle 2**: Find the sum of the terms in the following series:

$$100^2 - 99^2 + 98^2 - 97^2 + 96^2 - 95^2 + \cdots + 4^2 - 3^2 + 2^2 - 1^1$$

Write a script to sum the terms. Can you find the sum without Scilab or calculator?

```
// mathpuzzle2
```

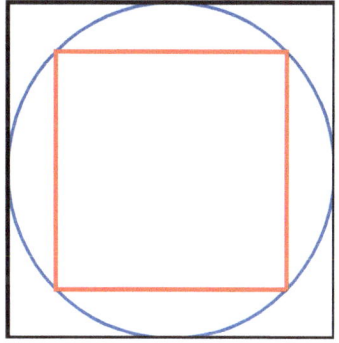

Figure 3.1: Inscribing circle

```
// Finding the sum of series
// 100^2-99^2+98^2-97^2 + ... + 2^2-1^1
// Construct two vectors u and :
u = 100:-2:2;
v = 99:-2:1;
w=u.^2-v.^2;
S = sum(w)
```

Note the sum is 5050.

3.2 Functions

The Scilab programming language is built around functions. A *function* is a file that can accept input arguments and return output arguments. The name of the file and the function should be the same. Functions operate on variables within their environment, separate from the environment, one can access at the Scilab command prompt.

Scilab has both built-in functions and user-defined functions.

Built-in Functions

Scilab has many built-in functions. These include *abs, sqrt, cos, sin, log, gamma, sum* and many more. For example,

```
--> cos(%pi) % One input
 ans  =
  - 1.
--> round(cos(0:%pi/2:2*%pi)) % vector input
 ans  =
    1.    0.  - 1.    0.    1.
--> format(5)
--> exp(1:4) // Exponential function
 ans  =
    2.72    7.39    20.1    54.6
--> log(exp(1:4)) // Natural log: ln as inverse of exp
 ans  =
    1.    2.    3.    4.
```

User-defined Functions

User-defined functions are created with optional extension .sce and with the extension .sci if the function is to be included in the user library. Each function must start with a function definition line that consists of:
- the word **function** in small letters.
- an input variable within parentheses and multiple inputs are separated by commas.
- a function name
- an output variable and multiple variables must be included within brackets and separated by commas and equality between the output and the name of the variable
- The body of the function
- last line must be **endfunction**

The general form of a function definition is

```
function [output1, output2, ...] = functionname( input1, ... )
\\Statements
endfunction
```

where the *functionname* must be the same as the file name: **functionname.sce or functionname.sci**

We will demonstrate the function constructions with several examples.

3.2.1 Develop a function *triple.sce* that triples the input. Test your function by typing: *triple(5)* and *triple([3:6]* at the prompt sign. save the file as triple.sce for future use.

```
// This function computes the triple values
function y=triple(x)
    y=3*x
endfunction
--> triple(5)
 ans  =
    15.
--> triple([3:6])
 ans  =
    9.    12.    15.    18.
```

Notice that the function works on vector input as desired.

3.2.2   1. Develop a function *power2.sce* which squares the numerical input. Test your function.
  2. Develop a function *powern.sce* which raises the numerical input to any desired power *n*. Test your function.

```
// This function computes the square of numerical values
function y=power2(x)
y= x.^2
endfunction

--> power2(15)
    225.
--> power2(-3:2)
    9.    4.    1.    0.    1.    4.
// This function computes the powers of numerical values
function y=powern(x,n)
y= x.^n
endfunction

--> format(5)
--> powern(5,3)
    125.
--> powern(-3:2,4)
    81.    16.    1.    0.    1.    16.
--> powern(1:6,1/2)
    1.    1.41    1.73    2.    2.24    2.45
```

3.2.3 Develop a function *func1.sce* that calculates the values of the following function and vectorize it.

$$f(x) = \frac{2x^4 - 4x^3 + 3x - 2}{x^2 - x + 3e^{-2x}}$$

Test your function by typing: $func1(3)$ and $func1([3, 1; -1, 0])$ at the prompt sign. save the file for future use.

```
// This function computes the values of a complex function
// Use the vectorize operations .*, .^ and ./ with variables
function output=func1(x)
    output=(2*x.^4-4*x.^3+3*x-2)./(x.^2-x+3*exp(-2*x))
endfunction
//
--> format(6)
--> func1(3)
  ans  =
      10.15
--> func1([3, 1; -1, 0])
  ans  =
      10.154  - 2.463
      0.041   - 0.667
```

3.2.4 Develop a function *trig.sce* that calculates the sine, cosine, and the tangent of an acute angle θ in a right triangle with given Opposite side and Adjacent side. Test your function for Adjacent side 12 cm and Opposite side 16 cm.

```
// This function requires two inputs and three outputs
// Order of inputs is important
// Notice the brackets used with outputs
function [sine, cosine, tangent] = trig(adj, opp)
    hyp = sqrt(adj^2+opp^2)
    sine = opp/hyp
    cosine = adj/hyp
    tangent = opp/adj
endfunction
//
--> format(6)
--> [sine, cosine, tangent] = trig(12, 16)
  tangent  =
      1.333
  cosine  =
      0.6
  sine  =
      0.8
\\ or
--> [s, c, t] = trig(12, 16)
  t  =
      1.333
  c  =
      0.6
  s  =
      0.8
```

3.2.5 Construct a Scilab function *cube.sce*, to compute the function

$$y = x^3$$

Test your function by typing:

```
cube(2)
cube(-2:4)
```

```
// This function computes the cube values
function y = cube(x)
    y = x.^3
endfunction
//
cube(2)
cube(-2:4)

--> cube(2)
  ans  =
     8.
--> cube(-2:4)
  ans  =
   - 8.   - 1.    0.    1.    8.    27.    64.
```

3.2.6 Develop a Scilab function to convert degree Fahrenheit to degree Celsius using the conversion formula

$$C = (F - 32) \times \frac{5}{9}$$

Create a table to use in the output for the Fahrenheit values $0 : 10 : 110$

```
// This function makes the conversion from Fahrenheit to Celsius
//
function Cel= Fah_Cel(Fah)
    Cel = (Fah - 32)*5/9
endfunction
//
format(5)
F = 0:10:110;
C = Fah_Cel(F);
Fahrenheit_Celsius = [ F',C']
--> Fahrenheit_Celsius = [ F',C']
  Fahrenheit_Celsius  =
     0.    - 17.78
     10.   - 12.22
     20.   - 6.67
     30.   - 1.11
     40.     4.44
```

```
50.      10.00
60.      15.56
70.      21.11
80.      26.67
90.      32.22
100.     37.78
110.     43.33
```

3.2.7 In the following problem, we will develop a function *disc.sce* to plot a disc (circle with its interior) for a specified center, radius, and color.

```
// This function paints a circle with given
// center (h,k) and radius r, and colors c:
// 1 black; 2 blue; 3 green; 4 cyan; 5 red, and so on

function disc(h,k,r,c)
    t=linspace(0,2*%pi,200);
    x=h+r*sin(t);
    y=k+r*cos(t);
    plot2d(0,0,-1,'010',' ',[-10,-10,10,10])
    set(gca(),"foreground",c);//apply the color
    xfpoly(x,y)// to plot and fill the circle
    isoview(gcf(), "on");
endfunction
```

3.2.8 Create a target illustration with ten colors.

```
clf
disc(0,0,10,1),disc(0,0,9,2),disc(0,0,8,3)
disc(0,0,7,4),disc(0,0,6,5),disc(0,0,5,6)
disc(0,0,4,7),disc(0,0,3,8),disc(0,0,2,9)
disc(0,0,1,19)
// In the next section, we use a simple loop to create this target
```

3.2.9 Run the script file multidiscs.sce to produce the circular design of figure 3.3

```
//multidiscs - A script to draw a circular design
//    -----------------------------------------------------
// inner discs
discplot(0, 0, 4, 2)
discplot(0, 0, 3, 4)
discplot(0, 0, 2, 1)
discplot(0, 0, 1, 5)
// outer discs
discplot(5, 0, 1, 5) ,discplot(5, 0, .8, 1) //left
discplot(5, 0, .6, 4),discplot(5, 0, .4, 2)
discplot(0, 5, 1, 5) ,discplot(0, 5, .8, 1) //top
```

Figure 3.2: Target(Bullseye)

```
discplot(0, 5, .6, 4),discplot(0, 5, .4, 2)
discplot(0, -5, 1, 5) ,discplot(0, -5, .8, 1) //bottom
discplot(0, -5, .6, 4),discplot(0, -5, .4, 2)
discplot(-5, 0, 1, 5) ,discplot(-5, 0, .8, 1) //right
discplot(-5, 0, .6, 4),discplot(-5, 0, .4, 2)
```

3.2.10 Look at figure 3.4 and compare the central two discs on left and right side. Which disc is larger: Central Left or Central Right?. This is a classic visual illusion.
Write a script file by calling the function discplot.sce to produce the figure 3.4.

3.3 Scilab time functions

Scilab has several built in utility functions to record time and duration of calculations.

1. timer() returns the total CPU time (in seconds) used by Scilab since the previous invoke of timer().
2. getdate() returns a vector with time information.
3. clock() returns a 6-element date vector containing the current date and time in decimal format: [year month day hour minute seconds]
4. etime(t2, t1) returns the time in seconds between different time points.
5. tic() and toc() Two functions work together to measure elapsed time.
6. date returns current date.
7. calendar(y,m) returns the m month calendar and the year y.
 Inspect these functions at the prompt command.

Figure 3.3: Multidiscs

Inspect the time Scilab functions at the prompt command.

```
//Calendar() returns calendar of current month and year
--> calendar()
//calendar(2018, 9) returns calendar of September, 2018

--> calendar(2018, 9)
  Sep 2018
    M       Tu      W       Th      F       Sat     Sun
    0.      0.      0.      0.      0.      1.      2.
    3.      4.      5.      6.      7.      8.      9.
    10.     11.     12.     13.     14.     15.     16.
    17.     18.     19.     20.     21.     22.     23.
    24.     25.     26.     27.     28.     29.     30.
    0.      0.      0.      0.      0.      0.      0.
//clock returns the current year, month, day,
// hour, minutes, and seconds
--> clock
```

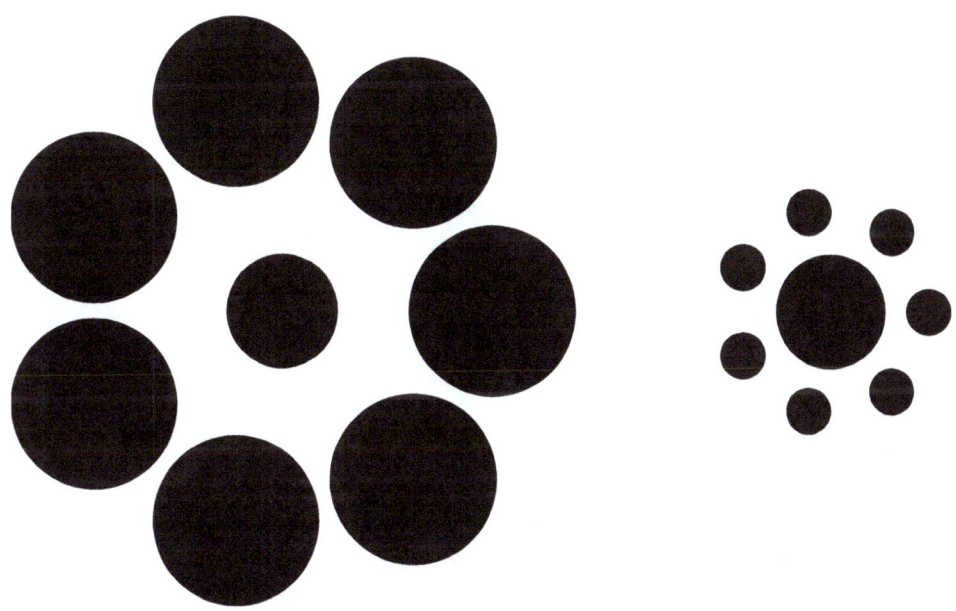

Figure 3.4: Circular illusion

```
// date or date() returns current day-month-year
--> date
--> date()
// getdate() returns a vector with current time information
--> current_time = getdate(); current_time'
```

3.3.2 Use the timer() function to compare the speed of your computer in processing simple calculations.

```
// This script uses timing functions to estimate the time taken to solve
5000 by 5000 system of linear equations.

tic();
timer();
rand('seed', 29);
n = 5000;// square dimension of a linear system
```

```
A = rand(n,n);
b = rand(n,1);
y=A\b;
cputime=timer()
toc()
//////
--> cputime=timer()
 cputime  =
     5.90625
--> toc()
 ans  =
     2.6070923
// Note two different time values.
```

3.4 Basic Scilab Programming

So far, computer programs (scripts and functions) we've constructed, every command or statement was executed in sequence. However, in many instances, executions of commands need not be in order or a selection of statements only. Sometimes, statements need to be repeated. Scilab provides many tools that can be used to control the flow of a program. These include logical statements, conditional statements using *if* and branching *elseif*. Loops are executed using *for* **and** *while*. In addition to many more built-in control functions.

3.4.1 Logical Operations

Comparisons between statements are true or false, and Scilab uses 1 for true and 0 for false. The relational and logical operators between two statements A and B are:

A > B	greater than
A >=	greater than or equal to
A < B	less than
A <= B	less than or equal to
A == B	equal to
A ~= B	not equal to
A & B	and
A ~ B	not
A | B	or

3.4.1 For the three numerical vectors created randomly as follows:

```
rand('seed', 131);
x = round(10*rand(1,7))
y = round(10*rand(1,7))
z = round(10*rand(1,7))
//
--> x = round(10*rand(1,7))
 x  =
     7.    8.    0.    9.    2.    5.    5.
--> y = round(10*rand(1,7))
```

```
 y =
    6.    7.    3.    1.    1.    2.    6.
--> z = round(10*rand(1,7))
 z =
    9.    2.    9.    10.    8.    4.    8.
```

Perform the following conditional statements:
1. $x > 5$ and $\sum(x > 5)$
2. $y >= 3 \& y < 8$
3. $x > y | x > z$

```
--> x>5
 ans =
  T T F T F F F
--> sum(x>5)  // How many numbers greater than 5
 ans =
    3.
--> y
 y =
    6.    7.    3.    1.    1.    2.    6.

--> y>=3 & y<8
 ans =
  T T T F F F T
--> x>y | x>z
 ans =
  T T F T T T F
```

find Function

The find function searches a matrix and finds the entries that satisfy a criterion. It returns a vector containing the indices of non-zero elements or the indices of true logical entries. test this function for the vector x from the previous problem:

```
--> x
 x =
    7.    8.    0.    9.    2.    5.    5.
--> find(x)
 ans =
    1.    2.    4.    5.    6.    7.
--> ind = find(x>5)
 ind =
    1.    2.    4.

--> x(ind)
 ans =
    7.    8.    9.
```

```
--> x(x>5) // same result as find
  ans   =
     7.     8.     9.
```

3.4.2 Selection Structures

Decisions in Scilab are made with **if** statement.

The if statement syntax is:

```
if (logical expression) then
    statements
end
```

If the logical expression is true then the statements are executed, otherwise they are skipped. For example,

```
r = 10;
if r > 0 then
    area = %pi * r^2;
end
--> area
area   =
    314.15927
```

The **else** statement allows to perform alternative statement whenever the logical expression is false. The if/else statement is

```
if (logical expression) then
    statements 1
else
  statements 2
end
```

For example, to simulate Head/Tail event, we use rand function to give a uniformly distributed random number between 0 and 1, and if the number greater than 0.5 we declare the result to be Head otherwise it is Tail:

```
if rand(1,1) > 0.5 then
    disp('Head')
  else
   disp('Tail')
end
```

3.4.2 Develop a Scilab function for the piece-wise function

$$f(x) = \begin{cases} -2x - 1 & ,x < -1 \\ \sqrt{1-x^2}+1 & ,-1 \leq x \leq 1 \\ 2x - 1 & ,x > 1 \end{cases}$$

Name the function pwise.sce and graph it on interval $[-2,2]$.

```
function y=pwise(x)
    if x > 1 then
        y= 2*x-1
    //elseif x >= -1& x<=1 // correct statement
      elseif x >= -1
        y=sqrt(1-x.^2)+1
    else
        y=-2*x-1
    end
endfunction
//
x = -2:0.01:2;
y=feval(x,pwise);
plot(x,y)
```

3.4.3 Develop a Scilab function to assign a letter grade for the score distribution:

$$0 \le F < 70 \le C < 80 \le B < 90 \le A \le 100$$

Test your function for the scores: 96, 35, 75, 88, 200.

```
// This function assigns letter grades to final scores
// The input must be a number
function y = grade(x)
    if x >= 0 & x <= 100 then
        if x >= 90 then
            y = 'A'
        elseif x >= 80
            y = 'B'
        elseif x >= 70
            y = 'C'
        else
            y = 'F'
        end
    else
        y = 'Input is not a real score'
    end
endfunction
// Check
--> grade(96), grade(64), grade(88), grade(200)
  ans  =
  A
  ans  =
  F
  ans  =
  B
  ans  =
  Input is not a real score
```

select and case Structure

The **select** statement executes groups of statements based on the value of a variable. The keywords **case** describe the groups. Only the first matching case is executed. There must be an **end** to match the **select**.

In the following example, we create a demo of a script file called top.sce and save it. Run the first option: Execute the options file with no echo from the menu bar. Test the script with different values.

```
// This script is a demo of select-case option
// top.sce
value = input('Enter a 1, 2, or 3 Proof: ');
select value
    case 1
    disp('It''s Gold!')
    case 2
    disp('It''s Silver!')
    case 3
    disp('It''s Bronze!')
    case 'Proof'
    disp('Here a Proof... BOOM!')
    else
    disp('Please read the instructions!')
end
```

3.4.3 Loop Control

A loop is a procedure to repeat the execution of a command or set of commands in sequence. In Scilab, The main two types of loops are *for-end* and *while-end*.

for loop

The general form of the **for** loop is: for loopvar = range action end

```
for variable = expression
    statements
end
// or
for variable = start: step : finish
statements
end
```

3.4.4 Use the for loop to print the first 4 perfect square and cube numbers. Use for loop the compute the area of a circle for $r = 1$ up to $r = 4$.

```
for k = 1:4
    [k;k^2;k^3]'
end
\\output
```

```
ans   =
   1.    1.    1.
ans   =
   2.    4.    8.
ans   =
   3.    9.    27.
ans   =
   4.    16.    64
for r=1:4
    area = 2* %pi * r^2;
    disp([r, area])
end
\\output
   1.    6.2831853
   2.    25.132741
   3.    56.548668
   4.    100.53096
```

3.4.5 Use for loop to compute the sum of the first 100 natural numbers

$$1+2+3+\cdots+98+99+100$$

```
sumk = 0;//initialize the sum
m = 4;   //show the first 4 steps
for k=1:m
    sumk = sumk + k;
    [k, sumk]  // print the first four steps
end
//
 ans   =
   1.    1.
 ans   =
   2.    3.
 ans   =
   3.    6.
 ans   =
   4.    10.
// Change m to 100
sumk = 0;//initialize the sum
m = 100;
for k=1:m
    sumk = sumk + k;
end
mprintf(' The total sum = %3.2f\n ', sumk )
// output
The total sum = 5050.00
```

3.4.6 Find the sum of the alternating harmonic series

$$\sum_{n=1}^{10^6}(-1)^{n-1}\frac{1}{n}$$

and compare this sum to $\ln 2$, and calculate the required cpu time on your machine.

```
timer()
sumk = 0;//initialize the sum
m = 10^6;    //show the first 4 steps
for k=1:m
    sumk = sumk + (-1)^(k-1)/k;
end
mprintf(' The total sum = %3.2f\n ', sumk )
diff = log(2)-sumk // The difference between the sum and the exact value
looptime = timer()
```

The output is

```
--> mprintf(' The total sum = %3.2f\n ', sumk )
 The total sum = 0.69
--> diff = log(2)-sumk
 diff  =
    0.0000005
--> looptime = timer()
 looptime  =
    1.25
```

It is important to avoid looping if possible, to accelerate the computations. In the next example, we will propose a vectorization scheme of the last problem and compare the ratio of looping time to vectorization time.

```
// vectorizing the loop
timer()
m = 10^6;
sk = 1: m;
//s = ((-1).^(sk-1)) ./ sk;
sumk = sum(((-1).^(sk-1)) ./ sk)
diff = log(2)-sumk
vectortime = timer()
looptime/vectortime
```

Note that the vectorization is about 13 times faster than looping calculation.

```
--> vectortime = timer()
 vectortime  =
    0.09375
--> looptime/vectortime
 ans  =
    13.333333
```

Nested Loops

Double and triple loops can be performed as in the next problem.

3.4.7 Compute the Hilbert matrix of size 4, using for loops. The entries in Hilbert matrix is given by the formula:

$$h(i, j) = \frac{1}{i + j - 1}$$

```
//  4 by 4 Hilbert Matrix
m=4; // number of rows/columns
for i = 1:m
    for j = 1:m
        H(i,j) = 1/(i+j-1);
    end
end
format(6)
H
spec(H)
```

The output is

```
H  =
    1.000    0.500    0.333    0.250
    0.500    0.333    0.250    0.200
    0.333    0.250    0.200    0.167
    0.250    0.200    0.167    0.143
```

while Loops

Scilab has the command while to use whenever you want to repeat the commands in a loop until another condition is met and without specifying the number of iterations. The format for a while loop is

```
while condition
statements
end
```

For example,

```
S=0;
while S < 10
    S=S+4
end
```

with output

```
S  =
    4.
 S  =
    8.
 S  =
    12.
```

3.4.8 Compute the number of scores between 75 and 95 in the following test scores:

$$67,78,89,94,45,98,82,56,90,75,31,83,97,80$$

```
test = [67, 78, 89, 94, 45, 98, 82, 56, 90, 75, 31, 83, 97, 80];
num = 0;
k = 0;
ntest = length(test);// number of scores in test
while k < ntest
    k = k+1;
    if test(k)>= 75 & test(k) <= 95
        num = num +1;
    end
end
num // print the result
```

the result is 8 scores.

3.4.9 A Fibonacci sequence is constructed of elements by adding the two previous numbers.

$$F_k = F_{k-1} + F_{k-2}, \qquad k = 3, 4, \cdots$$

The simplest Fibonacci sequences start with 0, 1 as follows

$$0, 1, 1, 2, 3, 5, 8, 13, \cdots$$

However, it may start with any two starting numbers.

Write a Scilab script Fibo.sce which computes the first 30 numbers and show that the ratio $F_k F_{k-1}$ approaches the golden ratio

$$\frac{F_{k-1}}{F_k} = \frac{\sqrt{5} - 1}{2} \approx 0.618$$

```
// Fibo.sce a script to compute
// Fibonacci sequence
m=12;
F(1) = 0; F(2) = 1;
for k = 3:m
    F(k) = F(k-1) + F(k-2);
end

plot((1:m-1)', F(1:m-1)./F(2:m), 'ro')
```

3.4.10 Develop a Scilab function bodymass.sce that calculates the body mass index (BMI) which is a measure of body fat based on height and weight that applies to adult men and women. The function asks the user for height and weight then prints the BMI and conclusions. Use one of the formulas

$$BMI = \frac{\text{Weight(in Kg)}}{\text{Height(in m)}} = \frac{\text{Weight(in lb)}}{\text{Height(in inches)}} \times 703$$

and use the following table for your recommendations.

BMI less than 18.5	Underweight (Risk of developing problems such as osteoporosis)
BMI from 18.5 to 25	Normal (healthy weight range)
BMI from 25 to 30	Overweight (Moderate risk of developing heart disease, diabetes)
BMI more than 30	Over-Overweight (High risk of developing heart disease, diabetes)

```
function    bodymass()
    H = input('Enter height in inches: ')
    W = input('Enter weight in lb: ')
    BMI = 703 * W / H^2
    mprintf(' Body Mass Index = %3.2f\n ', BMI )
    if BMI > 30 then
        disp('Over-Overweight (High risk of developing heart disease, diabetes)')
    elseif BMI > 25
        disp('Overweight (Moderate risk of developing heart disease, diabetes)')
    elseif BMI >18.5
        disp('Normal (healthy weight range)')
    else
        disp('Underweight (Risk of developing problems such as osteoporosis)')
    end
endfunction
```

At the prompt sign, test the function as follows

```
--> bodymass
Enter height in inches: 76
Enter weight in lb: 175
 Body Mass Index = 21.30
 Normal (healthy weight range)
\\ or
--> bodymass
Enter height in inches: 76
Enter weight in lb: 215
 Body Mass Index = 26.17
 Overweight (Moderate risk of developing
 health disease, diabetes)
```

3.4.11 Write an sci-file to plot the doubling phenomena that leads to chaos in logistic equation

$$x_{n+1} = rx_n(1 - x_n)$$

The relevant range of the parameter r is $[2.8, 4]$, as we increase r the number of equilibrium points doubles until it become chaotic. For further information refer to your differential equation textbook.

```
// resolution of bifurcation diagram
m = 200; n = 100;
// initial condition and parameter
x = rand(1,m); a = linspace(1,4,m);
// initialize variables
```

```
x = zeros(n,m);
x(1,:) = x0;
// step map forward
for i = 1:n
  x(i+1,:) = a.*x(i,:).*(1-x(i,:));
end
// plot
clf
plot(a,x(floor(2*n/3):n,:),'.k','markersize',.1);
xlabel('r'); ylabel('x')
```

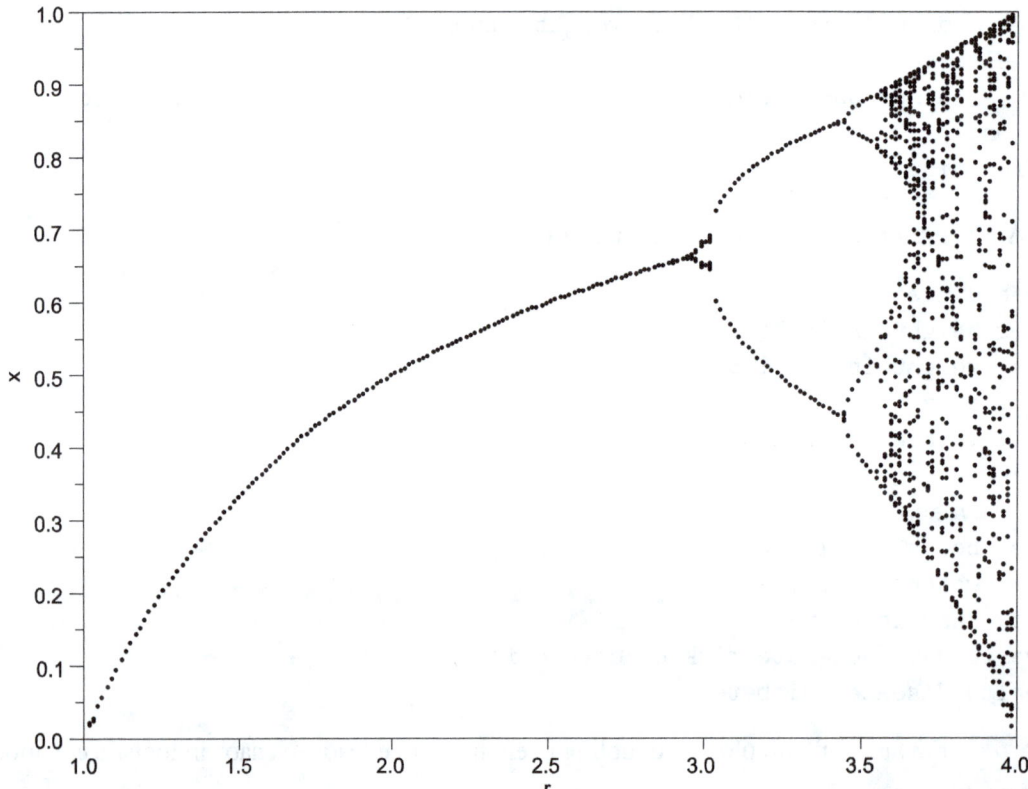

Figure 3.5: The Pitchfork diagram

3.4.12 Construct an sci-script to plot flowers in Figure 3.5.

```
// An SCRIPT-file flower.SCE script to produce
```

```
// "flower petal" plots
clf()
t = -%pi:0.01:%pi;
r(1,:) = 2 * sin(5 * t) .^ 2;
r(2,:) = cos(10 * t) .^ 3;
r(3,:) = sin(t) .^ 2;
r(4,:) = 5 * cos(3.5 * t) .^ 3;
for k = 1:4
  m=k;
    subplot(2,2,m)
polarplot(t, r(k,:))
end
```

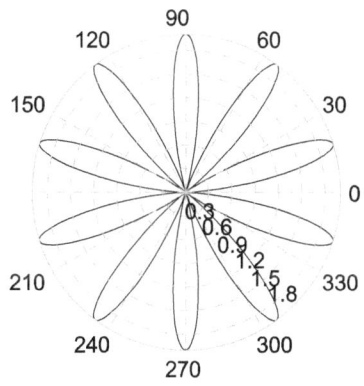

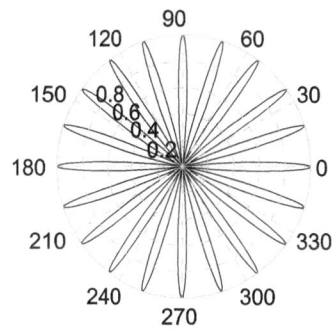

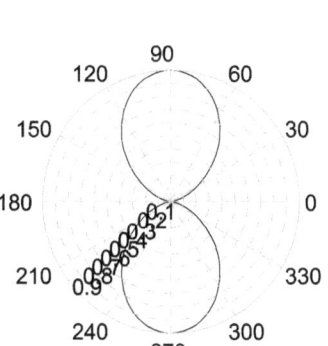

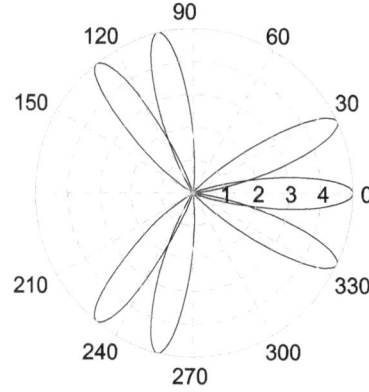

Figure 3.6: Polar flowers

3.4.13 Create a function that solves the quadratic equation

$$ax^2 + bx + c = 0.$$

Name the function `quadratic_eq.sce`. Make sure your function accepts vector input for a, b, c. It should be of the form: `[x1 x2]=quad2_eq(a,b,c)`. Test your function for
1. $a = 1, b = -5, c = 6$, `[s1, s2]=quad2_eq (1, -5, 6)`
2. $a = 2, b = -1, c = -1$, `[x1, x2]=quad2_eq (2, -1, -1)`
3. $a = [1,2,3], b = [-2,0,2], c = [1,1,-3]$: These represent three different quadratic equations.

```
// quadratic formula
function   [x1,x2] = quad2_eq(a,b,c);
           x1=(-b+sqrt(b.^2-4*a.*c))./(2*a);
       x2=(-b-sqrt(b.^2-4*a.*c))./(2*a);
endfunction // Run this script
-->[s1,s2]=quad2_eq(1,-5,6)
 s2  =
    2.
 s1  =
    3.
-->[x1, x2]=quad2_eq (2, -1, -1)
 x2  =
   - 0.5
 x1  =
    1.
-->a=[1, 2, 3], b=[ -2, 0, 2], c=[1 , 1, -3]
 a  =
       1.    2.    3.
 b  =
     - 2.    0.    2.
 c  =
       1.    1.  - 3.
-->[x1,x2]=quad2_eq(a,b,c)
 x2  =
       1.  - 0.70710i  - 1.38742
 x1  =
       1.    0.70710i    0.72075
```

3.4.14 Create a function sci-file that satisfies the condition for $-1 < x < 1$, $f(x) = 1$ and $f(x) = 0$ elsewhere. Name the function `cut_f (x)`. Test your function as follows:

```
// Cut Function
function out=cut_f( x);
   if  abs(x) < 1;
   out=1;
```

```
  else
   out=0;
   end
endfunction
 x=linspace(-%pi,%pi,200);
 y1=sin(2*x);
 for i=1:200
 y2(i)=sin(2*x(i)).*cut_f(x(i));
 end
 plot(x,y1,'b', x,y2,'r', 'thickness',3)
```

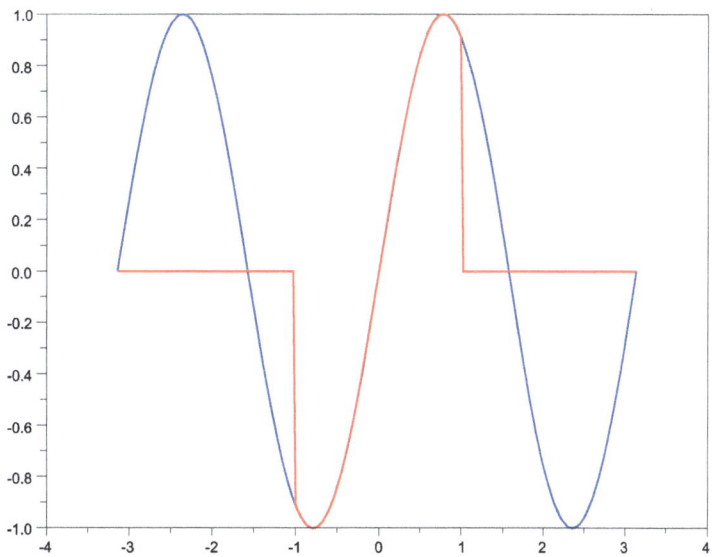

Figure 3.7: The Cut function

3.4.15 Find eight iterates of $x_{n+1} = 2x_n + 5, x_0 = -2$ and print them.

```
    x(1)=-2;
for i=2:7
    x(i)=2*x(i-1)+5;
end
x' // output in row format
```

3.4.16 Evaluate the following continued fraction

$$1 + \cfrac{1}{1 + \cfrac{1}{1 + \cfrac{1}{1 + \cfrac{1}{1 + \cfrac{1}{1+1}}}}}$$

Use the following script

```
    c=2;
for i=1:5
    c=1+1/c;
end
c
```

3.5 Homework: Programming, Functions and Scripts

3.5.1 Create functions to evaluate the following functions (select meaningful names):

1. $y(x) = x^3 - x$
2. $y(x) = \dfrac{1}{1 + x^2}$
3. $f(x) = cos(x^2)$

and check the functions for $x = 0, 1$

3.5.2 Create a function to change radians to degrees $y(x) = \dfrac{180}{\pi} x$ and check the function for $x = \pi/2, \pi/3$

3.5.3 Create a function to compute the area of the triangle $A(h, b) = \dfrac{hb}{2}$ and check the function for $h = 2, b = 10$

3.5.4 The harmonic series $\sum_{n=1}^{\infty} \dfrac{1}{n}$ is divergent. Write a script file to sum the series

1. $\sum_{n=1}^{100} \dfrac{1}{n}$
2. $\sum_{n=1}^{1000} \dfrac{1}{n}$
3. $\sum_{n=1}^{100000} \dfrac{1}{n}$

3.5.5 The series $S_n = \sum_{k=1}^{n} \dfrac{6}{k^2}$ is convergent. Find $\sqrt{S_n}$ for different values of n:

1. $n = 100$
2. $n = 1000$
3. $n = 10000$
4. $n = 100000$

Divide the above sums by the value π. What is your conclusion?

3.5.6 Evaluate the following continued fraction

$$1 + \cfrac{1}{1 + \cfrac{1}{2 + \cfrac{1}{3 + \cfrac{1}{4 + \cfrac{1}{5 + \cfrac{1}{6 + \cfrac{1}{7 + \cfrac{1}{8 + \cfrac{1}{9 + \cfrac{1}{10 + 1}}}}}}}}}}$$

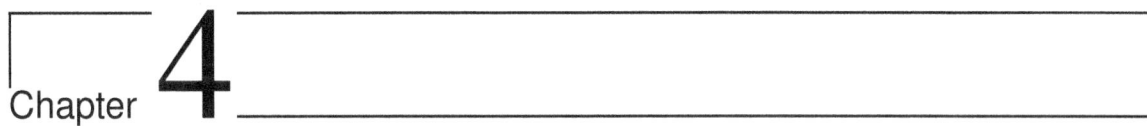

Chapter 4

2-D Graphics

Data visualization is an integral part of communicating ideas, decisions and in making complex information simpler to interpret and understand. The use of graphics is important for elementary to advanced students as well as scientists, engineers, and researchers.

Graphics are an essential part of the Scilab environment. Any newly encountered function should be plotted and analyzed. Understanding algebraic or differential equations using graphical techniques are very successful to understand the problems. Plotting and simulation the computational results are made simple with few Scilab commands. These commands are divided into two-dimensional, three-dimensional graphics, coloring techniques, and simulation. The two chapters on graphics will provide you with the necessary information, techniques, and examples to become a Scilab graphics expert.

4.1 Basic 2-D (Two-dimensional) Plots

Scilab provides extensive plotting capabilities to create a huge selection of graphs. The most common of all is the x-y plot for a set of ordered pairs of points given in a table of (x, y) values, or computed from a function $f(x)$ over an interval.

For example, to visualize the trend in the world population, we use the data given in the table

Year	Population(million)	Year	Population(million)
1900	1650	1960	3040
1910	1750	1970	3710
1920	1860	1980	4450
1930	2070	1990	5280
1940	2300	2000	6080
1950	2560	2010	6700

To plot these data we use the plot command as follows

```
x = 1900:10:2010;
y = [1650 ,1750 ,1860, 2070 ,2300 ,2560 ,3040 , 3710  , 4450,  ...
```

```
    5280 ,  6080,  6700];
```

```
plot(x,y)
```

A graphics window opens and we can dock it on Scilab window. We can add title, x- and y- labels and a grid by typing:

```
xgrid()
xtitle('The Growth of World Population', 'Year', 'Population (million)')
```

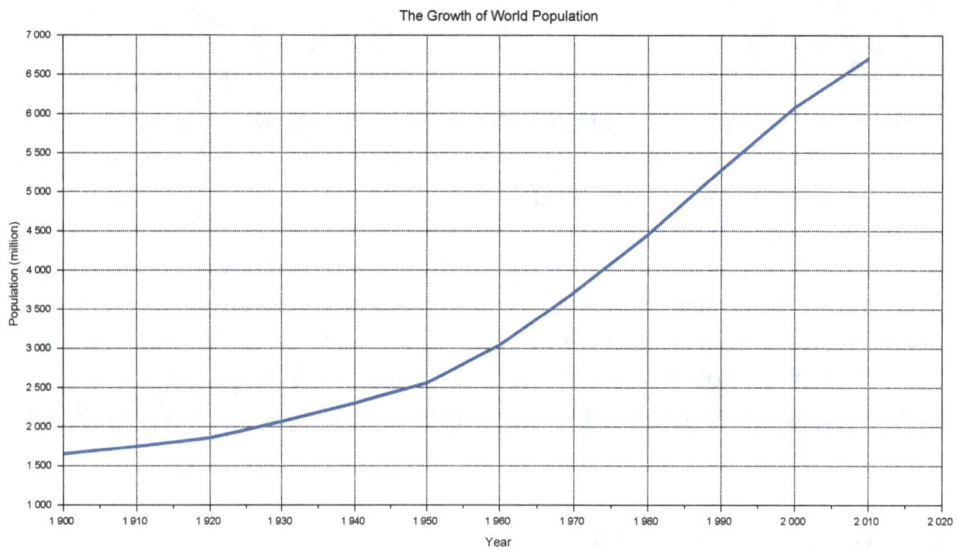

Figure 4.1: Simple plot of world population since 1900

The plot Function

The simple command *plot* is very versatile and has many options:

plot(y)	it will create a line of *y* versus the index numbers of *y*.
plot(A)	if *A* is a matrix, it will create multiple lines corresponding to each column.
plot(x,y)	it will display a graph of *y* versus *x*, provided *x* and *y* have the same size.
plot(x,A)	it will plot the matrix *A* versus the vector *x*. The sizes must match.

plot(A,x)	it will plot the vector x versus the matrix A. The sizes must match.
plot($\cdots$, 'str')	it will display according to the vectors, but using color, line type and different marker. These values are listed in the next table.
plot(x1,y1, 'str1', x2, y2, 'str2', $\cdots$)	it will plot $(x1, y1)$ with color and style according to str1, and it will plot $(x2, y2)$ with color and style according to str2, and so on.

In the following table, we will provide examples on how to implement the string options for given data x and y:

plot(x,y, '-.r')	it will create a red dash-dotted line.
plot(x,y, '–g')	it will create a green dashed line.
plot(x,y, 'o c')	it will create a cyan circle at each point.
plot(x1,y1, 'sb', x2, y2, 'pr')	it will create a line marked with blue square and a line marked with red star.

The string character in the plot command is used to change the line type, the color, and the symbol. The following table lists the characters that can be used:

Spec.	Line Style
'−'	solid line(default)
'− −'	dashed line
':'	dotted line
'−.'	dash-dotted line

Spec.	Color
'y'	Yellow
'r'	Red
'b'	Blue
'g'	Green
'k'	Black
'c'	Cyan
'm'	Magenta
'w'	White

Spec.	Marker Type
'.'	Point
'o'	Circle
'*'	Asterisk
'+'	Plus
'x'	Cross
's'	Square
'd'	Diamond
'>'	Right-pointing triangle
'<'	Left-pointing triangle
'^'	Upward-pointing triangle
'v'	Downward-pointing triangle
'p'	Pentagram(Five-pointed star)

Line and Symbol Thickness

The plot function has a string option to specify the width of the line and the character symbol in addition to their color:

plot(x,y, 'thickness', 3)	it will create a line with thickness 3 (default is 1).
plot(x,y, 'linewidth', 3)	it will create a line with thickness 3 (similar to Matlab).
plot(x,y, 'or', 'markersize',15),	it will create a red circle at each point with size 15.

Controlling the Graphics Window

Scilab creates the necessary objects to create and draw the graphics. All this information is stored in the graphics handles. Scilab will keep adding the new graphics objects to the current graphics window. To clear the old objects and draw the new graphics, use the command **clf** which clears the current graphics. Scilab has a command to create several graphics windows by specifying their id numbers **figure(num)**, these commands are summarized in the following table:

clf or clf()	it will clear current figure and reset it.
figure(2)	it will create a fugure with id 2
close()	it will close current figure.

4.1.1 Use Scilab to plot a sketch of a house and a car.

Solution: *We draw few simple lines to represent the house and the car and assign coordinates to few points, we use different colors and vary their thickness, we add wheels using the markers option as follows:*

```
clf // clear the old figures and duck the new figure
// Green House
xh = [1, 1, 4, 7, 7];
yh = [0, 4, 6, 4, 0];
plot(xh,yh, 'g','thickness',9)
// Cyan Door
xd = [3, 3, 5, 5];
yd = [0, 3, 3, 0];
plot(xd,yd,'c','thickness',7)
// Blue Car
xc = [9, 9, 11, 11, 12.5, 12.5];
yc = [0, 1.5, 1.5, 0.9, .7, 0];
plot(xc, yc,'b','thickness',3)
// Black Wheel with red tires
xw = [9.5, 12];  \\two points with circle marker
yw = [0.3, 0.3];
plot(xw, yw, 'o','markerfac','k','markeredg','r', 'markersiz',15)
```

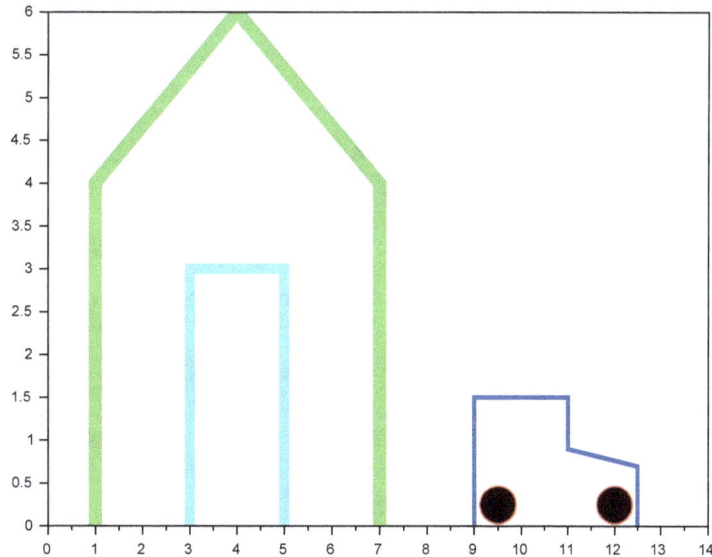

Figure 4.2: House and car

□

4.1.2 Plot the functions

1. $y = \sin(x)\sin(30x)$
2. $y = x\sin(5x) + 6$

3. $y = \dfrac{\sin(5x)}{x} - 6$

over the interval $[-2\pi, 2\pi]$, in separate figures, then plot them in one figure.

Solution:

```
// Construct the vectors
x = -2*%pi:0.0001:2*%pi; //x-coordinates with step size 0.0001
y1 = sin(x) .* sin(30*x);
y2 = x .* sin(5*x) +6;
y3 = sin(5*x) ./ x -6;
// individual plots
clf // clear the old figures and duck the new figure
plot(x,y1)
clf
plot(x,y2)
clf
plot(x,y3)
// To graph the three functions on the same figure
clf
plot(x,y1)// Scilab will add any graph to current figure
plot(x,y2)
plot(x,y3)
// Or we can graph them in one statement
clf
plot(x,y1,x,y2,x,y3)// Note Scilab color multiple plots
// Or we can put the three vectors in a matrix then plot the matrix
A = [y1; y2; y3]';
plot(x',A)
// We can graph A versus x easily (switching axes by shifting arguments)
plot(A,x')

//
```
□

Legends

The command **legend** creates a legend in the current plot and relates the line type with the text string that we pass in the order of creating the lines. In the following example, we use the legend command and coordinate shifts to separate curves.

4.1.3 Graph the functions
1. $y = x^2$
2. $y = 2|x|$
3. $y = 3e^{-x^2}$

over the interval $[-2.5, 2.5]$ with 17 points. Use different line styles, ten add legends to the plot.

Solution:

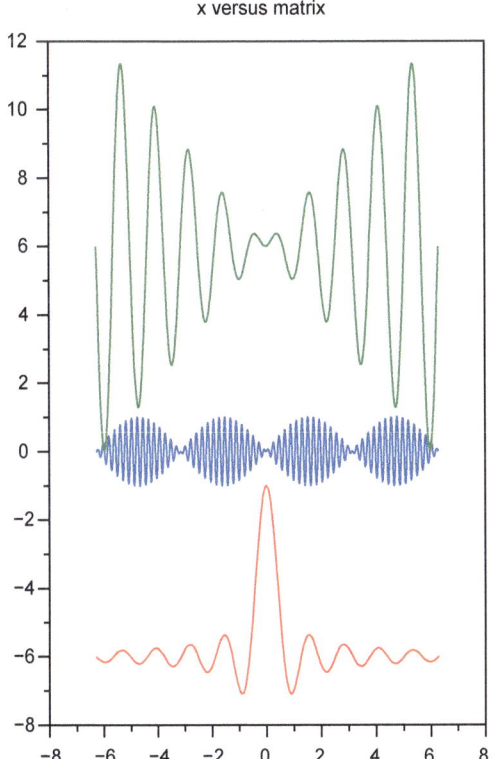 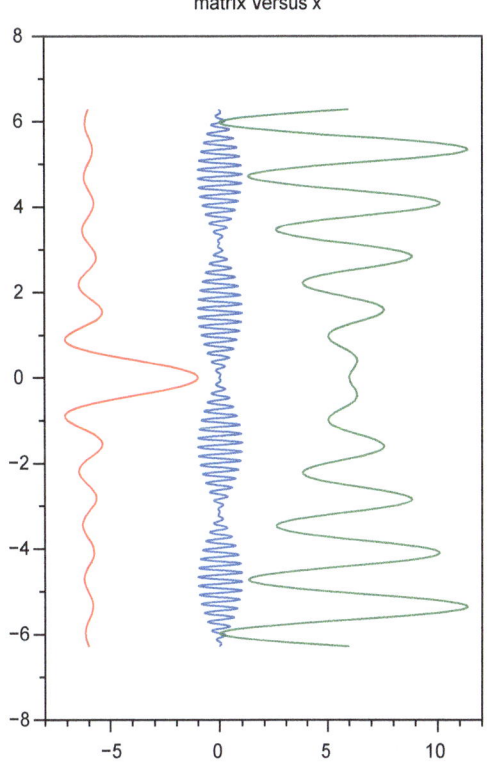

Figure 4.3: Multiple plots

```
x = linspace(-2.5,2.5,17);
y1 = x.^2;
y2 = 2*abs(x);
y3 = 3*exp(-x.^2);

clf
plot(x,y1,'--',x,y2,':',x,y3,'-.', 'thickness',2)
title('Without legends')

clf
plot(x,y1,'--',x,y2,':',x,y3,'-.', 'thickness',2)
legend(['$x^2$';'$2|x|$';'$3e^{-x^2}$']);
title('With legends')

//
```

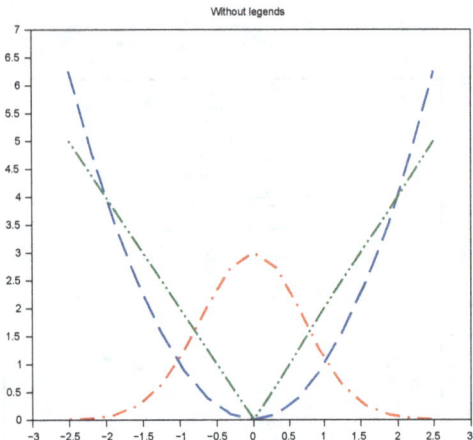
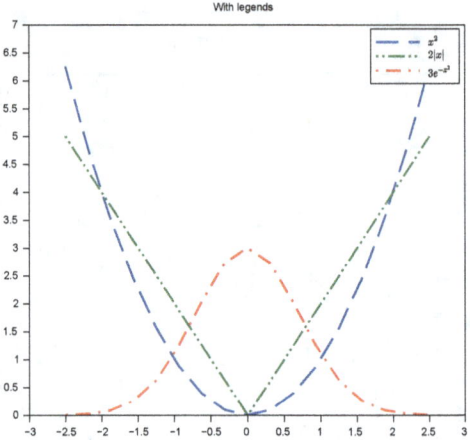

Figure 4.4: Legends and different line styles

4.1.4 Use vectors from the previous problem to plot the three functions next to each other with different markers and with their reflections.

Solution:

1. *Shift the parabola by six units to left, use a square marker. Its reflection is shifted up by four units using a circle marker.*
2. *The absolute value function is plotted with a diamond marker and its reflection is shifted up by four units using an asterisk marker.*
3. *The bell-shaped function is shifted to the right by five units using up triangle, and its reflection with lower triangle marker.*

```
clf
plot(x-6,y1,'s',x-6,-y1+4,'o', x,y2, 'd', x,4-y2,'*',..
    x+5, y3+2,'^',x+5, -y3+3,'v' )
title('Functions with different markers')
```

//
$\square$

4.1.5 Plot cubic function $y = x^3 - 2x$ over the interval $[-2, 2]$ using 21 points. Use the star marker and the circle marker with different colors and sizes.

Solution:

```
clf
x =linspace(-2,2,21);
y=x.^3 - 2*x;
```

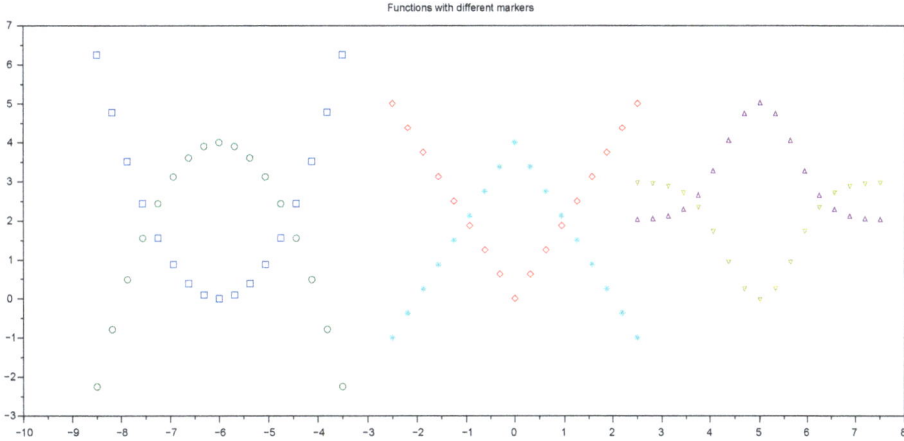

Figure 4.5: Shifted functions with different markers

```
plot(x,y, 'p','markerfac','g','markeredg','r', 'markersiz',20)
title('Cubic function in stars')
clf
plot(x,y, 'o','markerfac','y','markeredg','m', 'markersiz',20)
title('Cubic function in circles')
```

□

4.1.6 Scilab has many specialized plotting tools. Here we apply the rotation command to the previous cubic function to produce interesting water wheel.

Solution:

```
clf
x = -2:.02:2;
y=x.^3 - 2*x;
xy = [x;y]; //matrix format
for i=2*%pi*(0:10)/20,
  [xyr]=rotate(xy,i); // apply rotation
  xpoly(xyr(1,:),xyr(2,:),"lines") // use xpoly command
end
title('Flying wheel')
// Forcing isoview for all axes of the current figure
isoview(gcf(), "on")
```

□

Text and annotations in Scilab

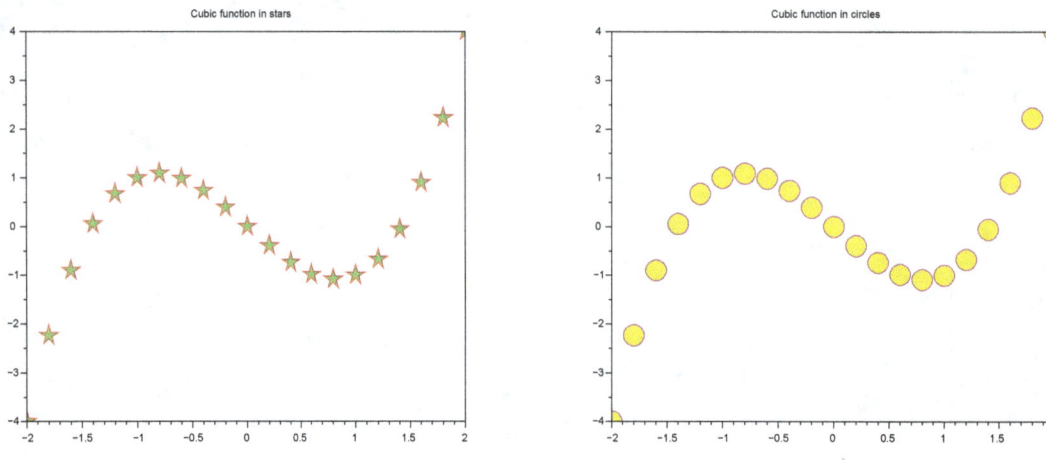

Figure 4.6: Cubic function in stars and circles

Scilab has many commands to include text in the plot. These include labels, titles, legends, and texts.

title('A title',<options>)	it will write the title centered at the top of the graph, the options include color, font size and more.
xtitle('title','xlabel','ylabel',<options>)	it will add title and labels to the axes with options.
legend('str1','str2', ...)	it will write a legend str1 and str2 as labels with options.
xstring(x,y,'txt', ...)	it will write a the string txt at the position (x,y) with options.

4.1.7 The function $p(t) = \dfrac{1}{1+20e^{-0.5t}}$ models the spread of rumors, where $p(t)$ is the proportion of the population that heard the rumor by the hour t. Plot this function and include labels and titles and draw a horizontal line at the 80% level to visualize the approximate hour at which this happens.

Solution:

```
clf
t = 0:0.1:15;
p = 1 ./ (1+20*exp(-0.5*t));
plot(t,p, 'linewidth', 2)
xtitle('Rumor Spread Model','time(hours)','proportion of population')
title('Rumor Spread Model','color','r','fontsize',3)// title has font option
xstring(8,0.6,'$ p(t) = \dfrac{1}{1+10e^{-0.5t}}$')
//Add the threshold at p=0.8
```

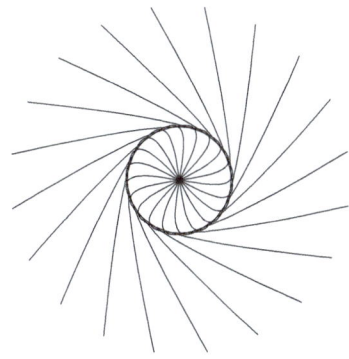

Figure 4.7: Water wheel

```
xsegs([0,15],[0.8,0.8]) // draws a segment connecting two points
xstring(1,0.8,['$80\%$'   'threshold'])
```

□

4.1.8 Plot the functions: $|\sin(x)|+1$, $x\sin(x)$, $\cos(1/x)-1$, $\cos(x)^5$ on $-\pi < x < \pi$. Use step size 0.02π. Use different colors and line types. Include legends.

Solution:

```
clf
x = -%pi:.02*%pi:%pi;
plot(x,abs(sin(x))+1,'-r');
plot(x,x.*sin(2*x),'--b');
plot(x,cos(1 ./(x))-1,':g');
plot(x,cos(x).^5,'-.k');
legend('$|\sin(x)|+1$','$x\sin(2x)$','$\cos(1/x)-1$','$\cos^5(x)$');
```

□

The same figure can be created by defining the y-coordinates as functions, then the plot command will call these functions to evaluate at x vector as follows:

```
clf
x = -%pi:.02*%pi:%pi;
function y=f1(x), y=abs(sin(x))+1; endfunction
function y=f2(x), y=x*sin(2*x); endfunction
function y=f3(x), y=cos(1/x) -1; endfunction
function y=f4(x), y=cos(x)^5; endfunction
plot(x,f1,'-r',x,f2,'--b',x,f3,':g',x,f4,'-.k')
legend('$|\sin(x)|+1$','$x\sin(2x)$','$\cos(1/x)-1$','$\cos^5(x)$');
```

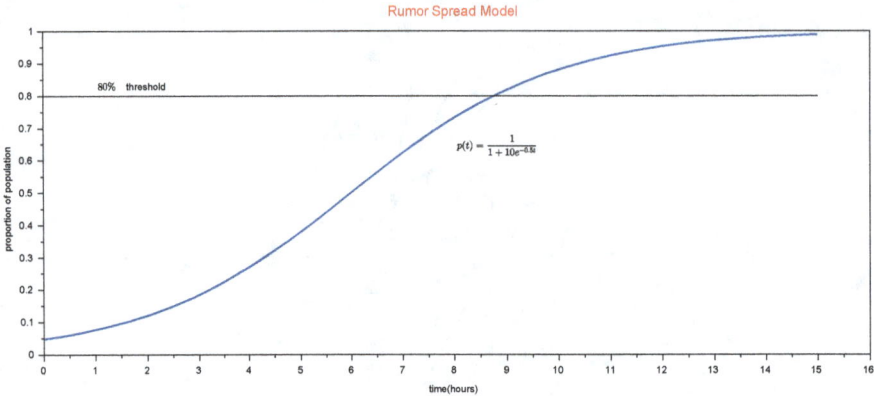

Figure 4.8: Modeling Rumor Spread

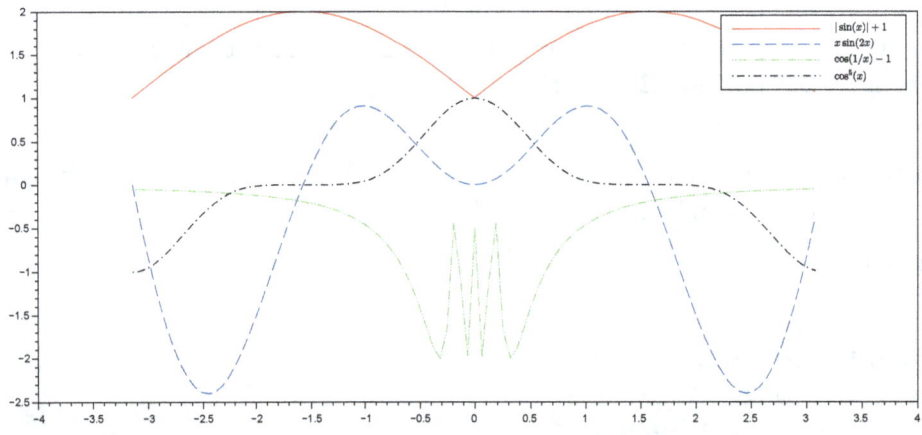

Figure 4.9: Function with legends

4.1.9 Plot the cardinal sine function (sinc)

$$y = \frac{sin(\pi x)}{\pi x}, \qquad x \neq 0$$

over the interval $[-25, 25]$

```
function y=fn(x)
    y= sin(%pi*x)/(%pi*x)
endfunction
```

```
x=-25:.02:25;
clf
plot(x,fn,'r','thickness',3) // Note the name of the function fn(x)
xtitle('Sinc function','x','y','fontsize',3)
xstring(5,0.5,'$y=\dfrac{\sin(\pi x)}{\pi x}$',0,1)
```

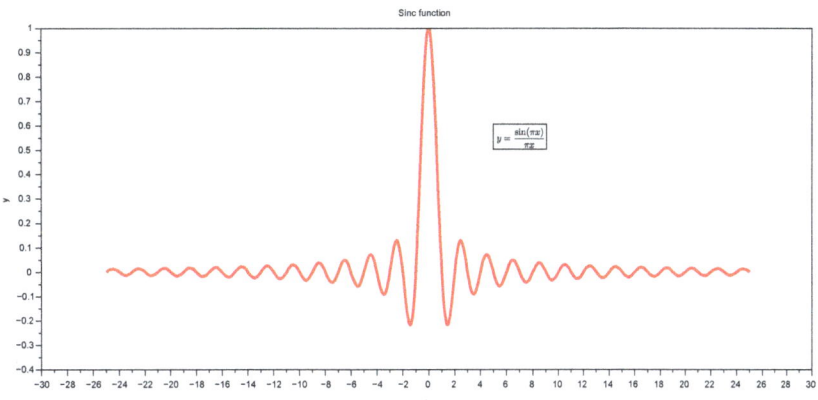

Figure 4.10:

4.1.10 Graph the function $y = (x-2)^2 \sqrt[3]{x(x+1)^2}$ on $[-2,3]$

```
function y=f(x)
    y= ((x-2)^(2)).*(x.*(x+1)^2)^(1/3)
endfunction
x=-2:.02:3.5;
clf
plot(x,f(x),'k','thickness',3)
title('$y=(x-2)^2\:\sqrt[3]{x(x+1)^2}$','fontsize',3)
```

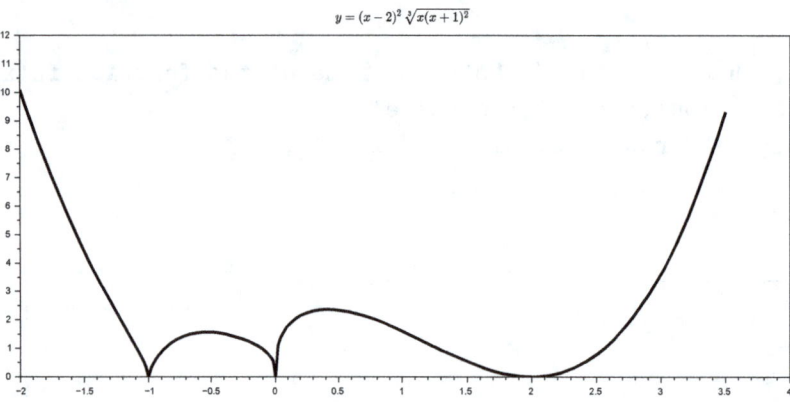

$$y = (x-2)^2 \sqrt[3]{x(x+1)^2}$$

Figure 4.11:

4.1.11 Combine the following increasing frequency functions into one matrix, then plot the matrix.
1. $y = \sin(x)$
2. $y = \sin(2x) + 2$
3. $y = \sin(4x) + 4$
4. $y = \sin(8x) + 6$

ob the interval $[0, 2\pi]$.

Solution:

```
clf
x = 0:0.05:2*%pi;
A = [sin(x);sin(2*x)+2;sin(4*x)+4;sin(8*x)+6];
plot(x,A,'thickness',3)
title('Functions with increasing frequencies','fontsize',5)
```

□

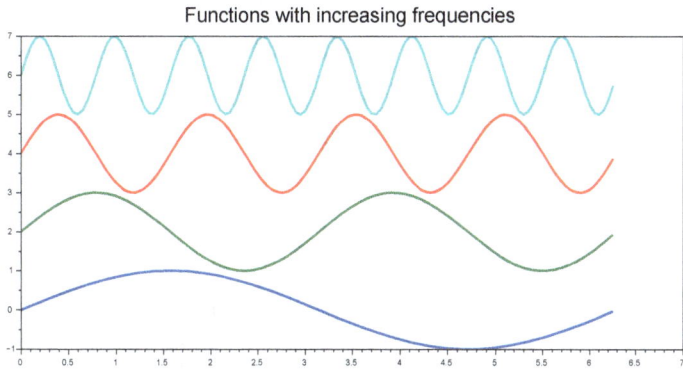

Figure 4.12: Matrix plot

4.1.1 Parametric and Polar Plots

parametric Equations

Curves can be described via parametric equations, by imagining a particle moving along the path as a function of time, thus the coordinates of the particle are functions of t:

$$x = x(t), \qquad y = y(t), \qquad a \leq t \leq b$$

so the initial position is the point $(x(a), y(a))$ and the final position is $(x(b), y(b))$.

Polar Equations

Functions in polar coordinates are by the equation

$$r = f(\theta), \qquad \alpha \leq \theta \leq \beta$$

where r represents the distance of a point from the origin as it rotates between the two angles α and β.

Functions in polar coordinates $r = f(\theta)$ can be written in parametric format as follows:

$$x = f(\theta)\cos(\theta), \qquad y = f(\theta)\sin(\theta), \qquad \alpha \leq \theta \leq \beta$$

or in t notation:

$$x = f(t)\cos(t), \qquad y = f(t)\sin(t), \qquad \alpha \leq t \leq \beta$$

Scilab graphical commands for parametric and polar equations:

plot(x,y)   it will plot the parametric equations and it has many options.

comet(x,y)	it will draw an animated plot of y versus x and it has color option.
paramfplot2d(f,x, t)	it will plot an animated curve defined by function $f(x)$ along parameter t.
polarplot(θ,r)	it will plot the radius r versus angle θ in radians and it has color option.

We will demonstrate these commands in the following examples.

4.1.12 Graph a unit circle inside an ellipse using parametric equations for the circle

$$x = \cos(t), \qquad y = \sin(t), \qquad [0, 2\pi]$$

and for the ellipse

$$x = 2\cos(t), \qquad y = \sin(t), \qquad [0, 2\pi]$$

use colors to identify them and use comet command to animate both.

Solution: *Use the following code and note the use of the new command **square**.*

```
t = 0:.01:2*%pi;
// Parametric Equations of the unit circle
x = cos(t);
y = sin(t);
clf
plot(x,y,'b', 'thickness',3)
// Parametric Equations of the ellipse
x = 2*cos(t);
y = sin(t);
plot(x,y,'r', 'thickness',3)
title('Unit circle and Ellipse')
// For correct perspective, use the square command
square(-2,-2,2,2)

// Comet Animation
clf
t = linspace(0,2*%pi,5000)';//increase number of points to slow animation
comet([cos(t),2*cos(t)],[sin(t),sin(t)],'colors',[9 5])
```

$\square$

4.1.13 Graph the limocon $r = 2\cos(\theta) - 1$ and the three circles $r = 2\cos(\theta), r = -2\cos(\theta), r = 1$ using the **polarplot** command.

Solution:

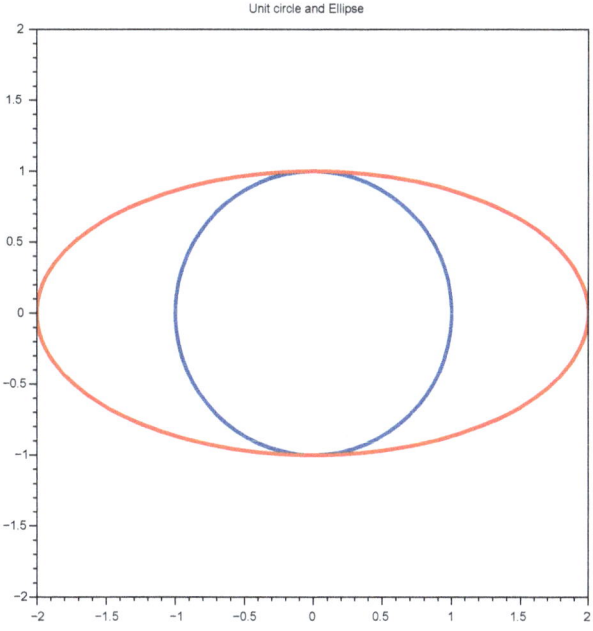

Figure 4.13: Unit Circle and Ellipse

```
t = 0:.01:2*%pi;
clf
polarplot(t,2*cos(1*t)-1)
title('Limasson r = 2cos(t)-1')

clf
t = 0:.01:2*%pi;
r=t*0+1;
polarplot([t' t' t'],[2*cos(t')   -2*cos(t') r'],[1,2,5])
title('Three Circle')
```

□

4.1.14 Graph the parametric equations

$$x(t) = t + 3|\sin(3t)|$$

$$y(t) = t + 3\cos(4t)$$

for $-\pi < t < \pi$. Use the two dimensional simulation command *comet* to follow the motion path.

```
t=-%pi:.001:%pi;
x=t+3*abs(sin(3*t));
```

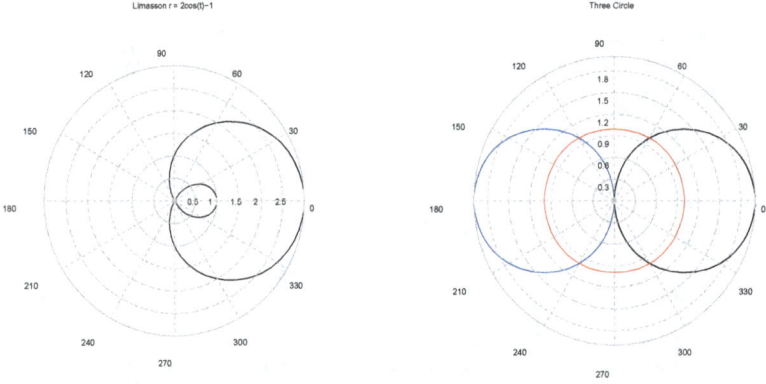

Figure 4.14: Functions in polar coordinates

```
y=t+3*cos(4*t);
plot(x,y,'b','thickness',3)
title('Parametric curve','fontsize',3)
\\ Comet Animation
clf
comet(x,y)
```

4.1.15 Plot the polar equation

$$r = \theta \sin(\theta)$$

for $-8\pi < \theta < 8\pi$

```
theta=-8*%pi:.01:8*%pi;
r=theta .* sin(theta);
polarplot(theta,r)
title('Polar curve','fontsize',3)
```

4.1.16 Graph the butterfly curve defined by the parametric equations:

$$x(t) = \sin(t)(e^{\cos(t)} - 2\cos(4t) - \sin(t/12)^5)$$

$$y(t) = \cos(t)(e^{\cos(t)} - 2\cos(4t) - \sin(t/12)^5)$$

for $0 < t < 12\pi$ and add eyes to it.

```
t=0:.01:12*%pi;
ct=cos(t);
p2=exp(ct)-2*cos(4*t)-sin(t/12).^5;
x=sin(t).*p2;
y=ct.*p2;
clf
```

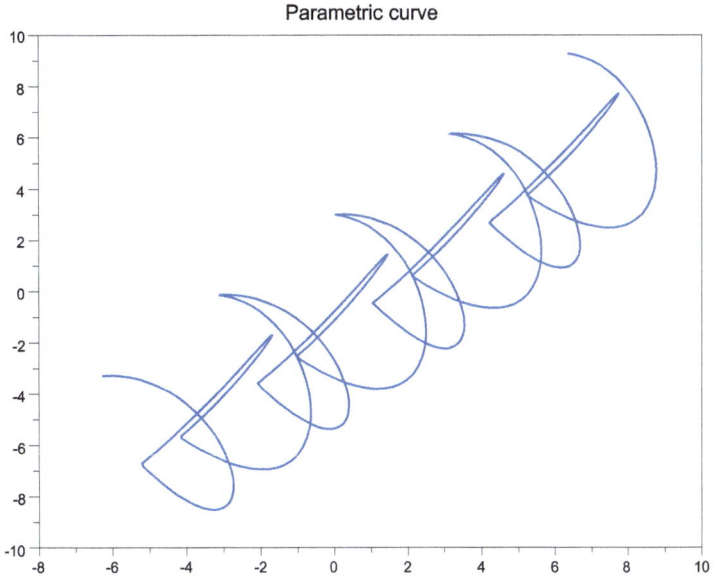

Figure 4.15: Parametric Curve

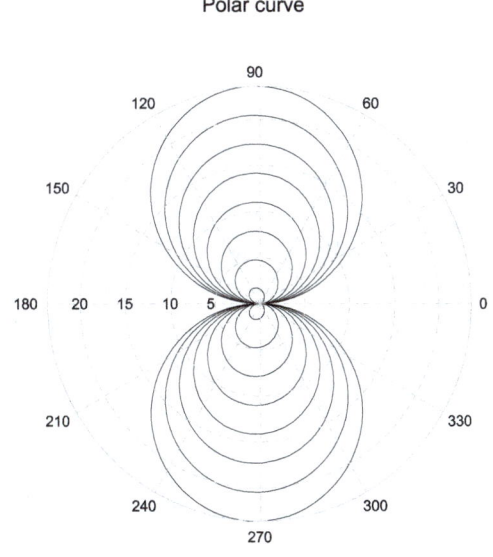

Figure 4.16: Polar Curve

```
plot(x,y)
xtitle('Butterfly Curve')
plot(0.3,1.7,'ro','markerface','r','markeredge','k','markersize',15)//right eye
plot(-0.3,1.7,'ro','markerface','r','markeredge','k','markersize',15)//left eye
```

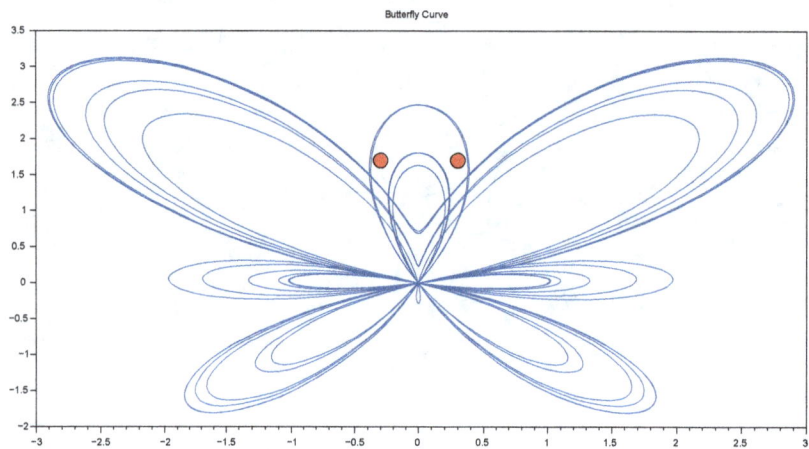

Figure 4.17: Butterfly Curve

4.1.2 Combining plots

It is possible to include several plots in a single graphics window. Scilab has the command subplot which divides the graphics window into an array of any desired smaller subregions. The command subplot(m, n, p) requires three numerical parameters, the first two numbers indicates the subdivision of the figure into m by n regions, and the third number to create the current region:

1. m: the first number identifies the number of row regions
2. n: the second number identifies the number of column regions
3. r: the third number identifies the location of the next plot which will be drawn. The counting for p is a row by row and left to right.

4.1.17 Plot the functions

$$f(x) = e^{-x^2} \sin(3nx)$$

where $n = 1, 2, \cdots, 6$ and over the interval $[-2, 2]$ into separate graphs on the same figure. Use 3 by 2 division.

```
clf
x= -2:.01:2;
// Use the loop to plot
 for p= 1:6
     subplot(3,2,p) // array 3 by 2 and location is p
     plot (x, exp(-x.^2 ).*sin(3*p*x),'k', 'thickness',2)
```

```
        title(" plot p = "+string(p),"fontsize",2);// identify the plot by its location
end
```

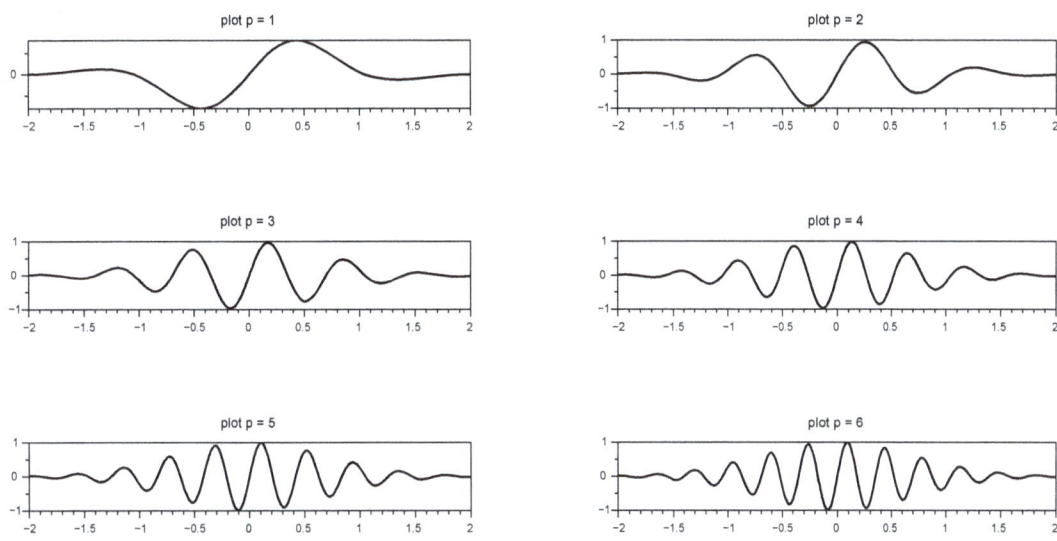

Figure 4.18: Sine packets

4.1.18 Plot the first nine Legendre polynomials on a 3 by 3 graphics window.

```
clf
x= -1:.01:1;
 for p= 1:9
     subplot(3,3,p)
     y = legendre(p-1,0,x);//built-in function
     plot (x, y,'k', 'thickness',2)
     title(" Legendre Polynomial n = "+string(p),"fontsize",2);
 end
```

4.1.19 Graph the the first six Bessel functions J_n, $0 \le n \le 5$ in a 2×3 figure.

```
clf
x= 0: 0.05 : 30;
 for n= 1:6
     subplot(2,3,n)
     plot (x, besselj(n-1, x),'k', 'thickness',3)
     title(" Bessel function n = "+string(n-1),"fontsize",2);
 end
```

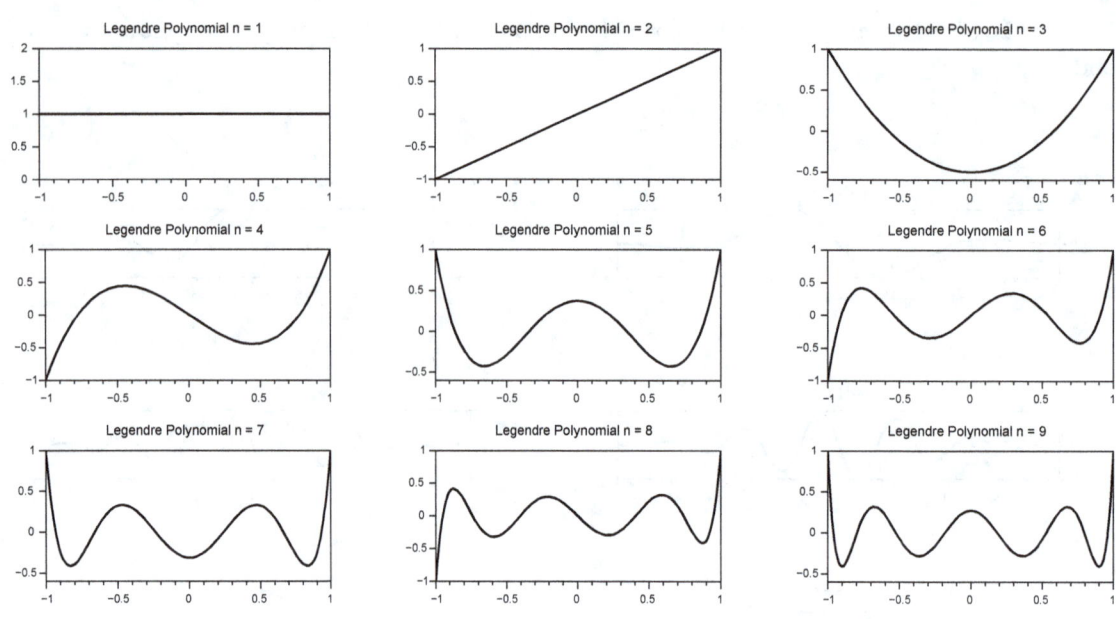

Figure 4.19: Legendre Polynomials

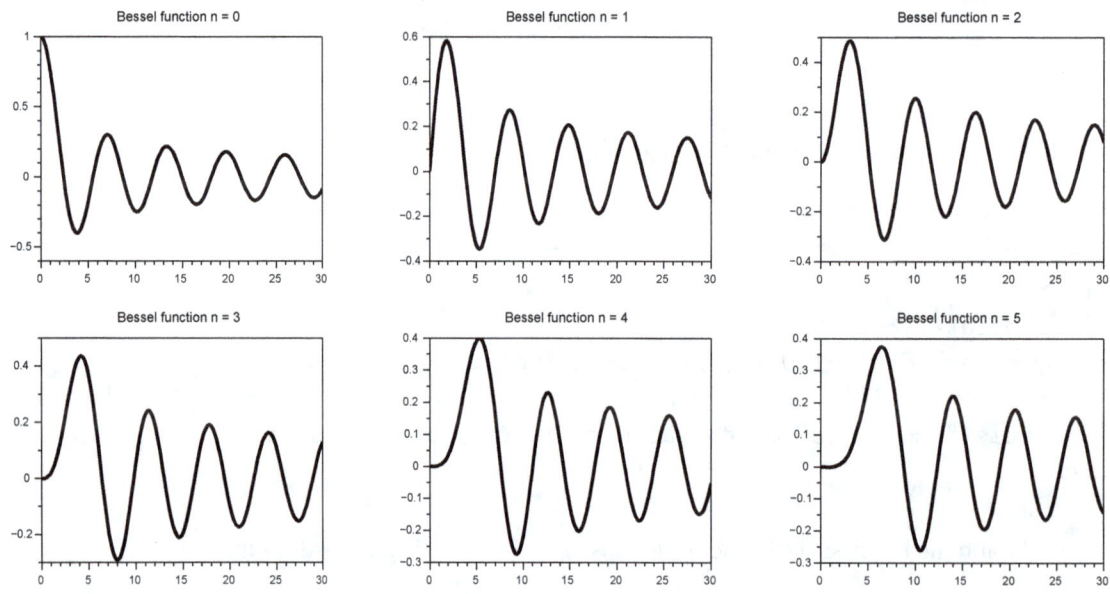

Figure 4.20: Bessel Functions

4.1.20 Divide the graphics window into two parts. IN the upper part, plot the function $y = x\sin(6x)$ on $[-6,6]$ and in the lower part plot the parabola $y = x^2 - 5$ using the *round* function over $[-3,3]$.

```
clf
x1=-6:.1:6;
x2=-3:.01:3;
subplot(2,1,1)
plot(x1,x1.*sin(6*x1),'thickness',3)
title('$x\sin(x)$','fontsize',3)
subplot(2,1,2)
plot(x2,round(x2.^2-5),'thickness',3)
title('$x^2-5$','fontsize',3)
```

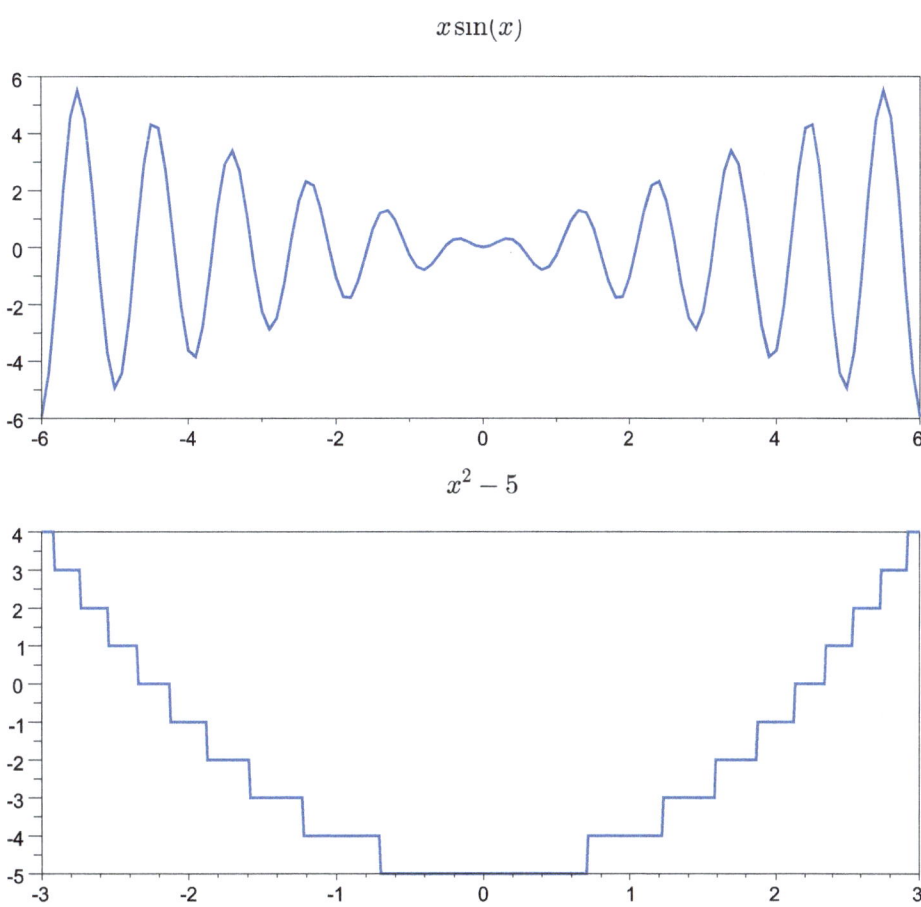

Figure 4.21: Two level figure

4.1.21 Plot the tie curve and the rose curve given in polar coordinates. Express both equations in parametric form and plot both curves as in the figure.

The tie curve is given by

$$r = 0.5 + \cos(2\theta), \qquad 0 \le \theta \le 2\pi$$

and its corresponding parametric equations are

$$x = r\cos(\theta)), \qquad y = r\sin(\theta)$$

and the rose curve is given by

$$r = \cos(16\theta/13), \qquad 0 \le \theta \le 26\pi$$

and its parametric equations are

$$x = r\cos(\theta)), \qquad y = r\sin(\theta)$$

Solution:

```
// Tie curve
clf
 t = 0:.01:2*%pi;
 r = .5 + cos(2*t);
subplot(2,2,1)
polarplot(t,r)
xtitle('Tie curve/polar')
subplot(2,2,2)
x=r.*cos(t);
y=r.*sin(t);
plot(x,y);
xtitle('Tie curve/parametric')
// Forcing isoview for all axes of the current figure
isoview(gcf(), "on")
// Rose curve
 t = 0:.01:26*%pi;
 r = cos(16*t/13);
subplot(2,2,3)
polarplot(t,r)
xtitle('Rose curve/polar')
subplot(2,2,4)
x=r.*cos(t);
y=r.*sin(t);
plot(x,y);
xtitle('Rose curve/parametric')
isoview(gcf(), "on")
```

□

4.1.22 Graph the functions
1. The double rose: $r = 1 - 2\sin(5t), \qquad 0 \le t \le 2\pi$
2. The Lituus: $r^2 = \dfrac{1}{t}, \qquad 0.1 \le t \le 10\pi$
3. Cosine Variation: $r = \cos(t/3) + \cos(t/2), \qquad 0 \le t \le 12\pi$
4. Golden Order: $r = \sin(99*t)(1.2 + sin^2(11*t)), \qquad 0 \le t \le 2\pi$

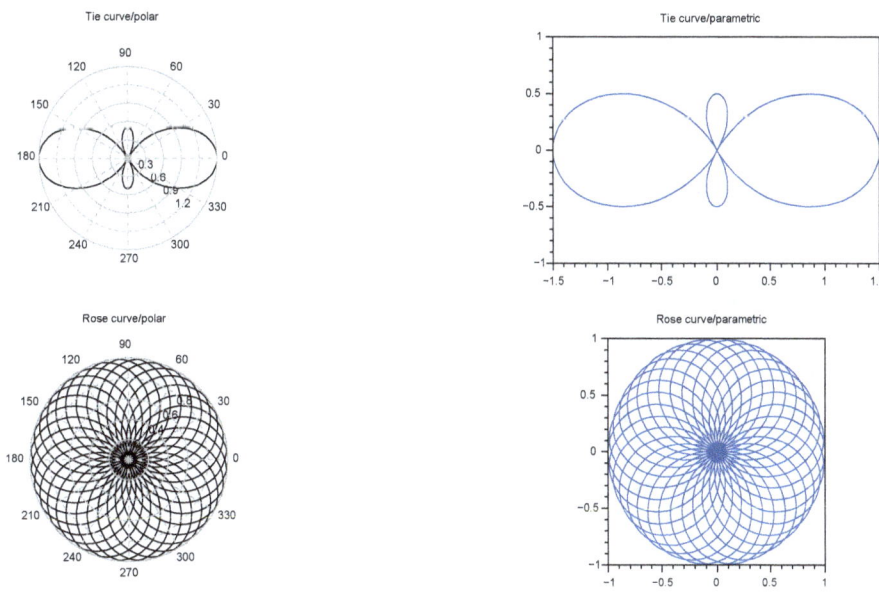

Figure 4.22: Polar/Parametric curves

Change to parametric equations then plot in 2 by 2 window and use the command **isoview**.

Solution:

```
//1. Double Rose
clf
t = 0.0:.01:2*%pi;
r=1-2*sin(5*t);
x=cos(t).*r;
y=sin(t).*r;
subplot(2,2,1)
plot(x,y)
title('Double Rose')
isoview(gcf(), "on")
//2. The Lituus
t = 0.1:.01:10*%pi;
subplot(2,2,2)
x=cos(t).*sqrt(1 ./t);
y=sin(t).*sqrt(1 ./t);
plot(x,y,-x,-y)
title('The Lituus')
isoview(gcf(), "on")
//3. Cosine Variation
t = 0:.01:12*%pi;
```

```
r=cos(t/3)+cos(t/2);
x=r.*cos(t);
y=r.*sin(t);
subplot(2,2,3)
plot(y,x,'r','thickness',2)
title('Cosine Variation')
isoview(gcf(), "on")
//4. Golden Order
t = 0:.001:2*%pi;
r=(1.2+sin(11*t).^2) .*(sin(99*t));
x=r.*cos(t);
y=r.*sin(t);
subplot(2,2,4)
plot(x,y,'y','markersize',1)
title('Golden Order')
isoview(gcf(), "on")
```

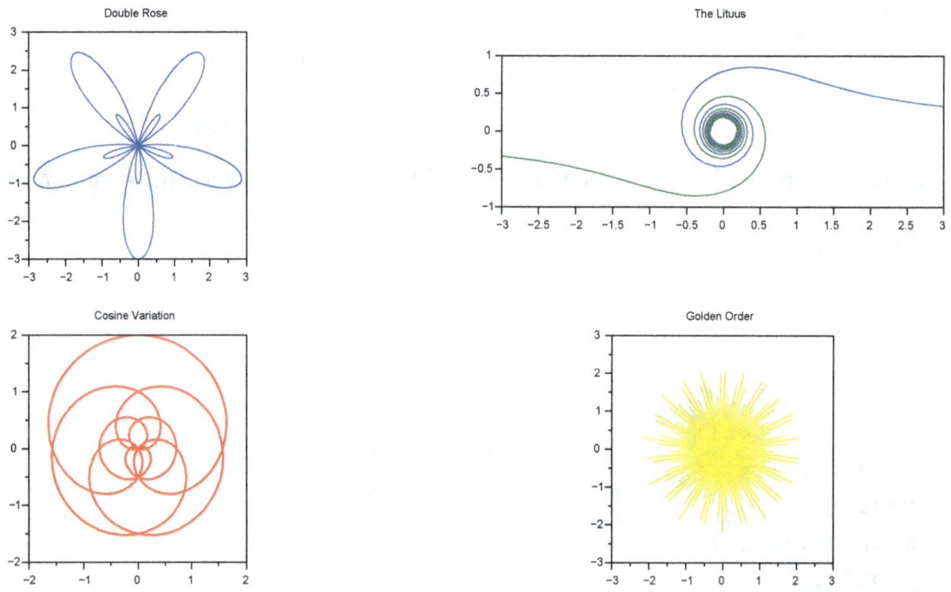

Figure 4.23: Parametric curves

4.2 Advanced 2-D Graphics

Scilab helps commands list many commands dedicated to 2d graphics which we will cover a few of these plotting commands. Also, Scilab allows the user to access every feature of the graphics window and its content through the so-called handle graphics which we will cover briefly.

Plotting commands

plot2d(. . .)   it will plot x versus y with many options:
- style, to control color by numerical values, style=2 for blue.
- strf, to control the display of captions, for example "010" removes box and axes around the figure.
- leg, to control legends, for example, " " omits the legends.
- rect, to control the viewing window.
- nax, to control the tics and labels on axes.

champ	it will a 2D vector field.
champ1	it will a colored 2D vector field.
histplot	it will draw a histogram.
bar	it will draw a bar graph.
grayplot	it will draw a 2D plot of a surface using colors.
scatter	it will draw a 2D scatter plot.
xfpoly	it will fill a polygon with color.
xsegs	it will draw segments between points.

For a complete list of graphics capabilities, consult the Scilab reference help manual.

Handle Graphics

Scilab allows users to access all graphics objects that can not be performed directly. There is a unique identifier (graphics handle) is associated with every graphics object. There are two main storage handles gca (for axes) and gcf (for figure) objects. We will demonstrate how to use them in the following example:

4.2.1 Plot the sinc function and make background gray and hide the box and the axes. Use its graphics handles.

```
x = 0:.2:6*%pi;
y = sinc(x);
clf
plot(x,y)
//
f = gcf;
f // to view all its options

--> f
 ans  =
```

```
The handle of type "Figure" with properties:
=======================================
children: "Axes"
figure_position = [-8,-8]
figure_size = [1072,656]
axes_size = [535,254]
auto_resize = "on"
viewport = [0,0]
figure_name = "Graphic window number %d"
figure_id = 0
info_message = ""
color_map = matrix 32x3
pixel_drawing_mode = "copy"
anti_aliasing = "off"
immediate_drawing = "on"
background =  -2
visible = "on"
rotation_style = "unary"
event_handler = ""
event_handler_enable = "off"
user_data = []
resizefcn = ""
closerequestfcn = ""
resize = "on"
toolbar = "figure"
toolbar_visible = "on"
menubar = "figure"
menubar_visible = "on"
infobar_visible = "on"
dockable = "on"
layout = "none"
layout_options = "OptNoLayout"
default_axes = "on"
icon = ""
tag = ""
//Try this command
f.background = -3; // changes the background to gray

// Second Graphics Handle
a = gca;
a

--> a
  ans  =
```

```
The handle of type "Axes" with properties:
========================================
parent: Figure
children: "Compound"

visible = "on"
axes_visible = ["on","on","on"]
axes_reverse = ["off","off","off"]
grid = [-1,-1]
grid_position = "background"
grid_thickness = [1,1]
grid_style = [3,3]
x_location = "bottom"
y_location = "left"
title: "Label"
x_label: "Label"
y_label: "Label"
z_label: "Label"
auto_ticks = ["on","on","on"]
x_ticks.locations = matrix 11x1
y_ticks.locations = [-0.4;-0.2;0;0.2;0.4;0.6;0.8;1]
z_ticks.locations = []
x_ticks.labels = matrix 11x1
y_ticks.labels = ["-0.4";"-0.2";"0";"0.2";"0.4";"0.6";"0.8";"1"]
z_ticks.labels = []
ticks_format = ["","",""]
box = "on"
filled = "on"
sub_ticks = [1,1]
font_style = 6
font_size = 1
font_color = -1
fractional_font = "off"

isoview = "off"
cube_scaling = "off"
view = "2d"
rotation_angles = [0,270]
log_flags = "nnn"
tight_limits = ["off","off","off"]
zoom_box = []
margins = [0.125,0.125,0.125,0.125]
auto_margins = "on"
axes_bounds = [0,0,1,1]

auto_clear = "off"
```

```
auto_scale = "on"

hidden_axis_color = 4
hiddencolor = 4
line_mode = "on"
line_style = 1
thickness = 1
mark_mode = "off"
mark_style = 0
mark_size_unit = "tabulated"
mark_size = 0
mark_foreground = -1
mark_background = -2
foreground = -1
background = -2
arc_drawing_method = "lines"
clip_state = "clipgrf"
clip_box = []
user_data = []
tag =

/// Try these commands

a.box = "off"
a.axes_visible = ["off","off","off"]
f.background = -2
```

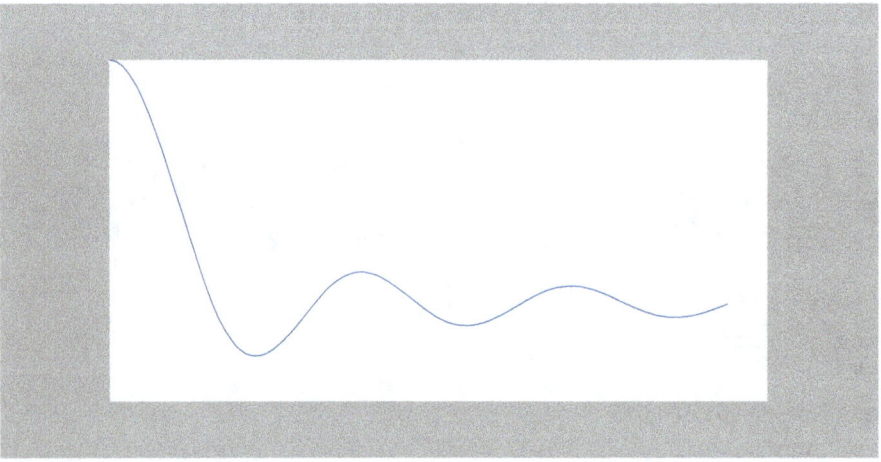

Figure 4.24: Plot manipulations using handles

4.2.2 Plot two red cardiods using the plot2d and xfpoly commands. The parametric equations of cardioid are

$$x(t) = (1 - \sin(t)) \cos(t)$$

$$y(t) = (1 - \sin(t))\sin(t)$$

for $-\pi < t < \pi$.

```
t=-%pi:.01:%pi;
x=(1-sin(t)).*cos(t);
y=(1-sin(t)).*sin(t);
plot2d(0,0,-1,'010','',[-2,-2.5,5,1.5]);
// '010' to hide axes.
// ''  for empty legends
xset('color',5)
xfpoly(x,y)//Fill the region with color
// Add another heart by shifting
xfpoly(x+2.5,y-.5)
```

Figure 4.25: Two Red Hearts

4.2.3 Draw the vector field of
- $(x, -y)$ on $(-1,1) \times (-1,1)$
- $(1, y - sin(x))$ on $(-3,3) \times (-3,3)$

Use champ1 and champ commands.

```
clf
x = linspace(-1,1,20);
y = linspace(-1,1,20);
[X,Y] = meshgrid(x,y);
```

```
subplot(121)
champ1(x,y,X',-Y',1.5,[-10,-10,10,10],"021")
title('Vector Field(x,-y)')
isoview(gcf(), "on")
x = linspace(-3,3,15);
y = linspace(-3,3,15);
[X,Y] = meshgrid(x,y);
subplot(122)
fy = Y-sin(X);
fx=ones(15,15);
champ(x,y,fx,fy',1.,[-10,-10,10,10],"021");
title('Vector Field(1,y-sin(x))')
```

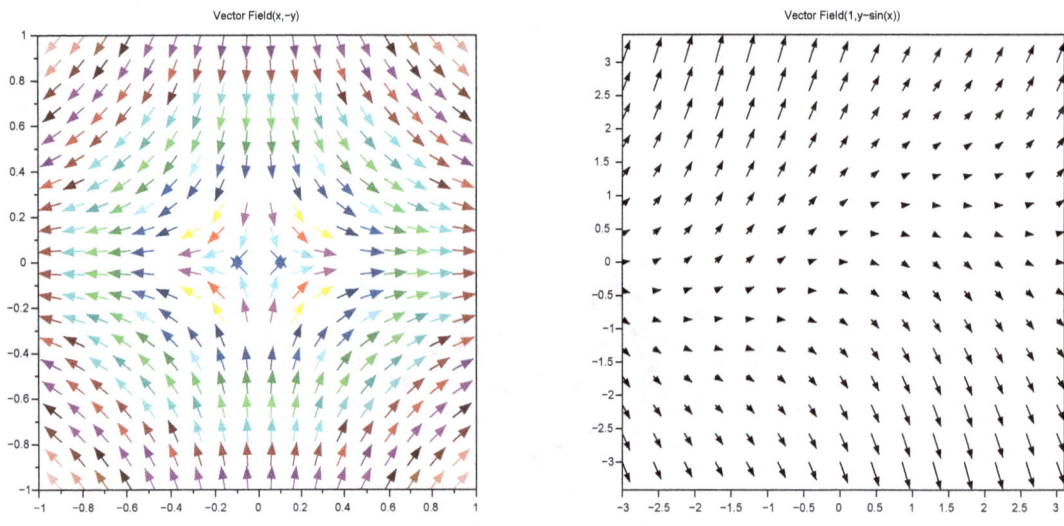

Figure 4.26: Vector fields

4.2.4 Graph a scatter plot for the father versus son height data of a small sample from Pearson data set.

```
//Enter the sample data set
father=[64.1,67.4,70.2,72.7,68.8,64.9,68.5,68.3,69.6,..
62.5,69.4,71.3,66.9,66.7,72.6,70.6,65.6,69.2,72.4,..
68.5,67.5,64.6,69.6,65.5,62.3,67.3,69.9,63,69.3,69,..
67.4,68.9,71.6,63.9,66.6,63.8];

son=[66.1,67.4,61.2,72.6,67.5,69.9,69.8,68.1,69.3,..
64,65.7,64.9,67,68.6,69.2,66.9,66.8,67.5,71.1,68.2,..
68.4,66.1,68.3,68.8,63.9,66.4,70.3,64.8,68.2,..
```

```
64.8,69.1,72.4,70.2,67.7,68.7,63.7];

scatter(father,son,"fill")
xlabel('father')
ylabel('son')
title('Pearson s Height Data')
```

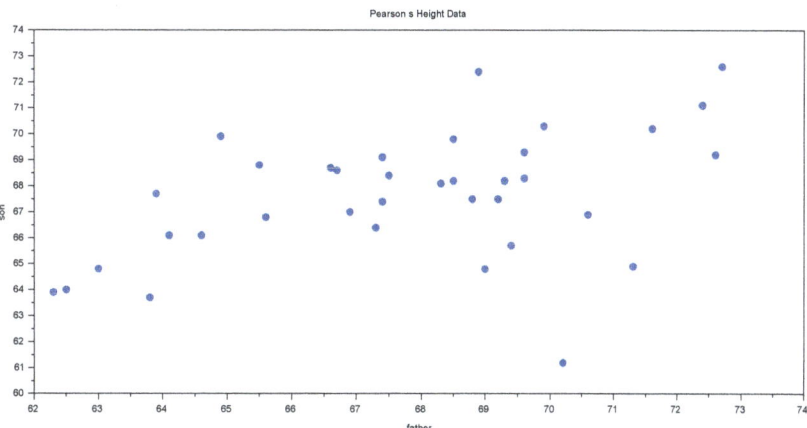

Figure 4.27: Pearson Height Data

4.2.5 Graph the cubic function and its derivative
1. $y = 3x - x^3$
2. $y = 3 - x^2$
on $[-3, 3]$.
- Use step size 0.1.
- Add a boxed title and labels to your plot.
- Add grid.
- Change viewing window to (-3,3) by (-5,5).
- Add system of coordinates at the origin.
- use isoview to show the correct perspective.

Solution:

```
clf
x = -3:0.1:3;
y = 3*x-(x).^3;// cubic function
plot(x,y, 'thickness',3);
dy = 3 - 3 * x.^2;// derivative function
plot(x,dy,'r', 'thickness',3);// plot in red
```

```
// Add title and labels
xtitle('Cubic function and its derivative' , 'X', 'Y', boxed =1 );
//Add grid
xgrid(4,2,8); // 4 for color, 2 for thickness, 8 for style
// Change viewing window to (-3,3) by (-5,5)
a=gca;
a.data_bounds=[-3,-5;3,5];
// Add system of coordinates at the origin
xsegs([0,0], [-5,5], 13); // Draw a segment as y-axis in green
xsegs([-3,3], [0,0], 13); // Draw a segment as x-axis in green
isoview(gcf(), "on")
```

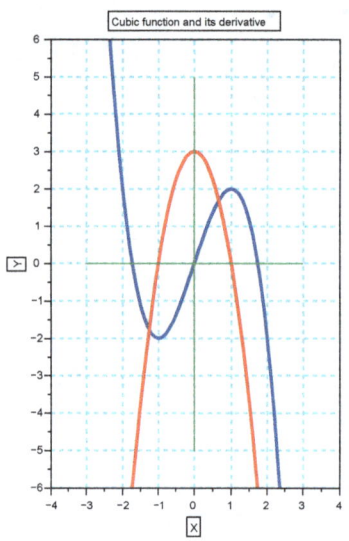

Figure 4.28: plot commands example

$\square$

4.2.6 Apply the following rigid transformations to the function $f(x) = 3x - x^3$:

- $f(x+6)$, left horizontal shift.
- $f(x-6)$, right horizontal shift.
- $f(x)+7$, positive vertical shift.
- $f(x)-7$, negative vertical shift.
- $-f(x)$, reflection.

Plot the function $f(x)$ and its transformations over interval $[-3,2]$, select viewing window $(-10,10), (-10,10)$, add legends and title.

Solution:

```
clf
x = -3:0.1:2;
y = 3*x-(x).^3;
plot(x,y,'r', 'thickness',6)
plot(x+6,y,'g', 'thickness',3)
plot(x-6,y,'b', 'thickness',3)
plot(x,y+7,'c',x,y-7,'m', 'thickness',3)
plot(x,-y,'k', 'thickness',3)
//axis([-8 8 -8 8])
//plot(y,x,'y', 'thickness',3)
a=gca;//get the handle of the newly created axes
a.data_bounds=[-10,-10;10,10]; // viweing window
title('Rigid Transformations','fontsize',3)
legend(['$f(x)$';'$f(x+6)$';'$f(x-6)$';'$f(x)+7$';'$f(x)-7$';'$-f(x)$']);
```

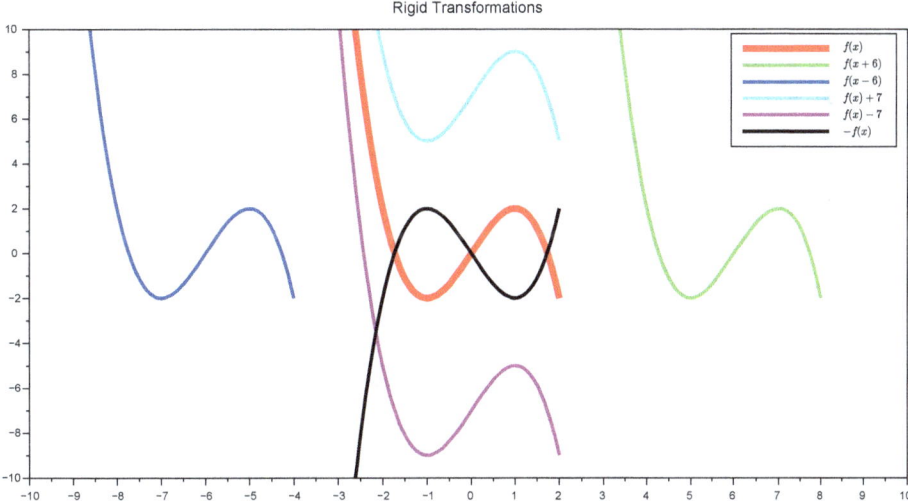

Figure 4.29: Function transformations

☐

4.2.7 In this example, we show how to draw basic geometric shapes: triangle, square, and circles and how to combine their plots.

Solution:

```
// Construct the x and y  vector coordinates
clf
//square
```

```
x_s = [-3  3 3 -3 -3];
y_s = [-3 -3 3  3 -3];
//triangle
x_t = 3*cos([%pi/2 %pi/2+2*%pi/3 %pi/2+4*%pi/3 %pi/2]);
y_t = 3*sin([%pi/2 %pi/2+2*%pi/3 %pi/2+4*%pi/3 %pi/2]);
//circles (more points)
x_c = 3*cos([0:%pi/50:2*%pi]);//outer circle
y_c = 3*sin([0:%pi/50:2*%pi]);
x_cc = (3)/2*cos([0:%pi/50:2*%pi]);//inner circle
y_cc = (3)/2*sin([0:%pi/50:2*%pi]);
//Plot the triangle in the upper left pane
subplot(2,2,1)
plot(x_t,y_t,'b','thickness',3);
a=gca;
a.axes_visible=["off","off","off"];
a.box ="off";
isoview(gcf(), "on");
xtitle('Triangle');
//Plot the square in the upper right pane
subplot(2,2,2)
plot(x_s,y_s,'r','thickness',3);
a=gca;
a.axes_visible=["off","off","off"];
a.box ="off";
isoview(gcf(), "on");
xtitle('Square');

//Plot the circles in the lower left pane
subplot(2,2,3)
plot(x_c,y_c,'g','thickness',3);
plot(x_cc,y_cc,'k','thickness',3)
a=gca;
a.axes_visible=["off","off","off"];
a.box ="off";
isoview(-3,3,-3,3);
xtitle('Circles');

//Plot all figures  in the lower right pane
subplot(2,2,4)
plot(x_s,y_s,'r','thickness',3);
plot(x_c,y_c,'g','thickness',3);
plot(x_t,y_t,'b','thickness',3);
plot(x_cc,y_cc,'k','thickness',3);
a=gca;
a.axes_visible=["off","off","off"];
a.box ="off";
```

```
isoview(gcf(), "on");
xtitle('Combining shapes');
```

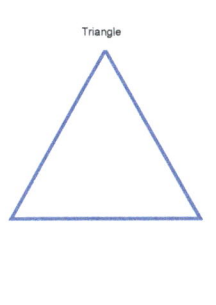

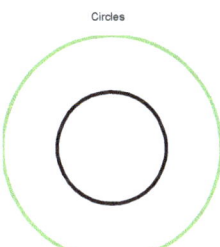
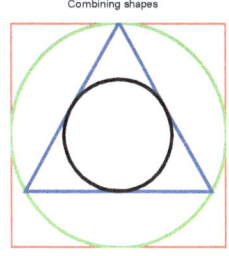

Figure 4.30: Example 4

☐

4.2.8 **Coloring geometric shapes**. In this example, we show how to construct various geometric shapes and filling them with colors.

Solution:

```
// Geometric shapes
clf
x_3=cos([0:2*%pi/3:2*%pi]);//triangle: 2pi/3
y_3=sin([0:2*%pi/3:2*%pi]);
x_4=cos([0:%pi/2:2*%pi]);//square: pi/2
y_4=sin([0:%pi/2:2*%pi]);

x_5=cos([0:2*%pi/5:2*%pi]);//pentagon: 2pi/5
y_5=sin([0:2*%pi/5:2*%pi]);
x_6=cos([0:2*%pi/6:2*%pi]); //hexagon: 2pi/6
y_6=sin([0:2*%pi/6:2*%pi]);
x_7=cos([0:2*%pi/7:2*%pi]); //septagon: 2pi/7
y_7=sin([0:2*%pi/7:2*%pi]);
x_8=cos([0:2*%pi/8:2*%pi]); //octagon: 2pi/8
y_8=sin([0:2*%pi/8:2*%pi]);
```

```
x=0:.01:2*%pi;
x_c=cos(x); //circle
y_c=sin(x);
x_w=0:.02:6;y_w=exp(-.4*x_w).*sin(4*x_w); // wavy function

//Creating shapes with colors on figure 20 by 20
//Shift the coordinates to arrange shape on the figure

plot2d(0,0,-1,"010"," ",[0,0,20,20])
xset("color",2)
xfpoly(4+2*x_3, 16+2*y_3);
xset("color",5)
xfpoly(10+2*x_4, 16+2*y_4);
xset("color",3)
xfpoly(16+3*x_5, 16+3*y_5);
xset("color",7)
xfpoly(4+2*x_6, 10+2*y_6);
xset("color",6)
xfpoly(10+2*x_7, 10+2*y_7);
xset("color",18)
xfpoly(16+3*x_8, 10+3*y_8);
xset("color",13)
xfpoly(4+2*x_c, 4+2*y_c);
xset("color",19)
xfpoly(10+1*x_c, 4+2*y_c);
xset("color",1)
xfpoly(16+3*x_c, 3.5+3*y_c);
xset("color",22)
xfpoly(16+3*x_6, 3.5+3*y_6);
xset("color",32)
xfpoly(16+3*x_4, 3.5+3*y_4);
xtitle('Colorful geometric figures')
```

$\square$

4.2.9 Mandelbrot Set, Plot the Mandelbrot set.

Mandelbrot set M is a set of complex numbers, $z = a + bi$, for a, b real, for which the sequence z, $f_z(z) = z^2 + z$ and its iterations $f_z^n(z)$ does not tend to ∞ as $n \to \infty$.

Solution: *We will plot two versions of Mandelbrot set, adapted from Matlab codes.*

```
//Mandelbrot version1
niter = 50;
gridSize = 400;
n=gridSize;
xlim = [-2, 1];
```

Colorful geometric figures

Figure 4.31: Example 6

```
ylim = [ -1.5,  1.5];
x = linspace( xlim(1), xlim(2), gridSize );
y = linspace( ylim(1), ylim(2), gridSize );
[xGrid,yGrid] = meshgrid( x, y );
```

```
c = xGrid + %i*yGrid;
z = zeros(n,n);
k = zeros(n,n);
// Calculate
for ii = 1:niter
    z   = z.^2 + c;
    k(abs(z) > 2 & k == 0) = niter - ii;
end
// Show
clf
xset("colormap",jetcolormap(64))//try hotcolormap or others
Sgrayplot(x,y,k)
a=gca;
a.box = "off";
a.axes_visible = ["off","off","off"];
title('Mandelbrot 1')
// Mandelbrot version 2
maxIterations = 200;
gridSize = 400;
n=gridSize;
xlim = [-0.748766713922161, -0.748766707771757];
ylim = [ 0.123640844894862,  0.123640851045266];
x = linspace( xlim(1), xlim(2), gridSize );
y = linspace( ylim(1), ylim(2), gridSize );
[xGrid,yGrid] = meshgrid( x, y );
z0 = xGrid + %i*yGrid;
count = ones(n,n );
// Calculate
z = z0;
for n = 0:maxIterations
    z = z.*z + z0;
    inside = abs( z )<=2;
    count = count + inside;
end
count = log( count );
// Show
clf
xset("colormap",jetcolormap(64))
Sgrayplot(x,y,count, strf="050")
a=gca;
a.box = "off";
a.axes_visible = ["off","off","off"];
title('Mandelbrot 2')
```

□

4.3 Homework: 2-d curves

4.3.1 Use an increment of 0.02 for the interval $0 \leq x \leq 2\pi$ then plot the function $y = \sin(15x)\sin(3x)$. Add title and labels to the plot

4.3.2 1. Subdivide a figure into two rows and one column
  2. In the top row, plot $y = \tanh(2x)$ on the interval $-4 \leq x \leq 4$ in increments of 0.1.
  3. Add a title and labels to the plot
  4. In the lower row, plot x versus $y = \tan(x)$ on the same range
  5. Add a title and labels to the plot

4.3.3 Subdivide a figure into four regions and plot one function in each subplot with titles and as follows
  1. $y1 = \sin(x)$ in red and solid
  2. $y2 = \sin(2x)$ in blue and dashed
  3. $y3 = \sqrt[3]{3x - x^3}$ in green and dotted
  4. $y4 = 9x^2 - x^4$ in yellow and star
on interval $[-\pi, \pi]$ in increments of 0.1

4.3.4 Plot the functions on the same system of coordinates
  1. the Gauss curve $y = e^{-x^2/2}$
  2. the Cauchy curve $y = -\dfrac{2}{1+x^2}$
for the interval $[-3, 3]$ in increments of 0.02

4.3.5 **Parametric curves**: Subdivide a figure into four regions and plot one curve in each region, include titles
  1. $x = 3\sin^5 t, \quad y = 3\cos^5 t$ for $0 \leq t \leq 2\pi$
  2. $x = 9\cos t - 3\cos 4t, \quad y = 9\cos t - 3\sin 4t$ for $0 \leq t \leq 2\pi$
  3. $x = 3t - 2\sin t, \quad y = 3 - 2\cos t$ for $-8 \leq t \leq 8$
  4. $x = 2t - 3\sin t, \quad y = 2 - 3\cos t$ for $-8 \leq t \leq 8$

4.3.6 **Curves in polar coordinates**: Subdivide a figure into two regions and plot one curve in each region, include titles
  1. $r = 2\sin^2 \theta \tan^2 \theta$ for $-\pi/3 \leq \theta \leq \pi/3$
  2. $r = \dfrac{4}{1 + \sin^2 \theta}$ for $0 \leq \theta \leq 2\pi$

Chapter 5

3-D Curves

Drawing three-dimensional graphs of data sets can lead to a deeper understanding of the patterns and trends of the data. Scilab provides a comprehensive set of 3-dimensional graphical tools. Some built-in functions plot lines in three dimensions, others plot surfaces, and slices. Colors are used to represent different configurations and to represent the fourth dimension. Thus, manipulating the colors is essential to seeing complicated shapes in three and four-dimensional functions. In this chapter, the basic tools and techniques needed in producing three-dimensional plots are introduced with useful examples. Each problem serves as a model to graph a different type of data to gain a better understanding.

5.1 Line Plots

There are three plotting functions which can be used to draw three dimensional curves:

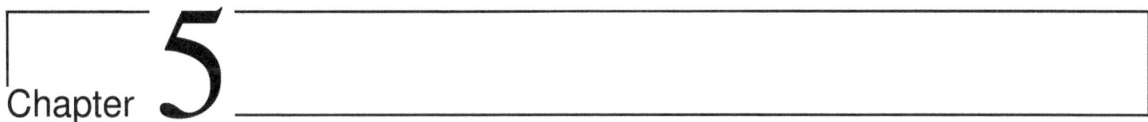

param3d(x,y,z, . . .)	plots 3-d curves, if x, y, z are vectors of the same length.
param3d1(x,y,z, . . .)	plots 3-d multiple curves, if x, y, z are vectors of the same length.
comet3d(x,y,z)	draws animated 3-d curves.

Three dimensional curves defined parametrically by three coordinate equations:

$$x = f(t)$$
$$y = g(t)$$
$$z = h(t)$$

for a parameter $t \in [a, b]$. The graphing of three dimensional line plots involves:

1. Creating a parametric vector t over the interval $[a, b]$.
2. Creating the corresponding three vectors x(t), y(t), z(t) from the given formula.
3. Using param3d(x , y, z) and annotating the curve.

---

4. Using comet3d(x , y, z) to simulate the motion of a point (object) moving along the curve.

5.1.1 Draw the circular helix

$$x = \cos(t)$$
$$y = \sin(t)$$
$$z = t$$

over the interval $[0, 20]$. Annotate your plot.

Solution:

```
//Circular Helix
clf
t =linspace(0,30,1000);
x=cos(t);
y=sin(t);
z=t;
// plot the curve without options
subplot(2,2,1)
param3d(x,y,z)
//Add viweing angles and labels
subplot(2,2,2)
param3d(x,y,z,45,15,"cos(t) @ sin(t) @ t")
//Add title and color red and thickness
subplot(2,2,3)
param3d(x,y,z,45,15,"cos(t) @ sin(t) @ t")
title('Circular Helix')
e=gce(); //the handle on the 3D polyline
e.foreground=color('red');
e.thickness=3;
//Draw two springs next to each other
subplot(2,2,4)
param3d(x,y,z,45,15)
e=gce();
e.foreground=color('green');
e.thickness=3;
param3d(x+3,y+4,1.5*z,15,5)
e=gce();
e.foreground=color('magenta');
e.thickness=2;
title('Two Springs')
```
□

5.1.2 Draw the parametric spherical curve

$$x = \cos(32 * t)\cos(t)$$
$$y = \sin(32 * t)\cos(t)$$
$$z = t$$

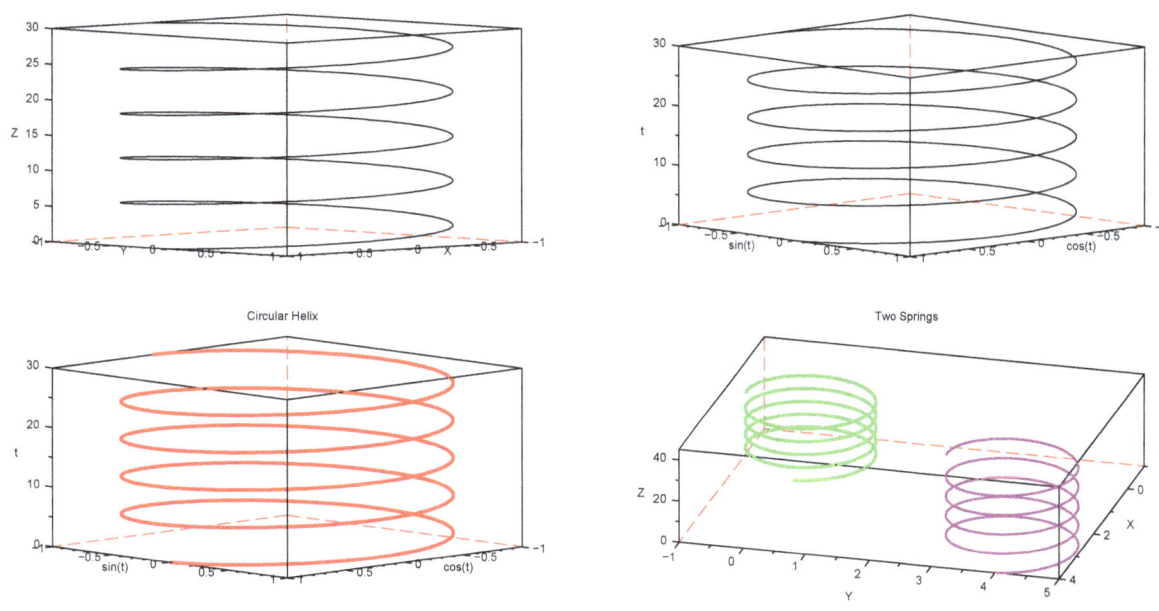

Figure 5.1: Springs

over the interval $[-pi/2, \pi/2]$. Use the handle files to remove the axes and the box, add color and thickness and title. Animate this curve.

Solution:

```
//Spherical curve
t=linspace(-%pi/2, %pi/2, 2000);
x=cos(32*t).*cos(t);
y=sin(32*t).*cos(t);
z=sin(t);
param3d(x,y,z,25,55,"X@Y@Z",[6,4])
//
e=gce(); //the handle on the 3D polyline
e.foreground=23;
a=gca(); //the handle on the axes
a.box="off";// omit boxing
a.axes_visible = ["off","off","off"];//omit axes
//omit axes labels
a.x_label.text =" ";
a.y_label.text =" ";
a.z_label.text =" ";
title("Spherical curve")
// Animation
```

```
clf
comet3d(x, y, z)
```

□

Spherical curve

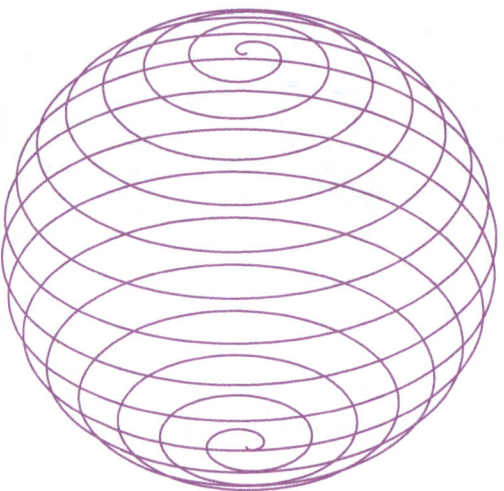

Figure 5.2: Spherical curve

5.1.3 Draw the spiral Bed spring helix

$$x = (5 + |t - 5\pi|)\cos(2t)$$
$$y = (5 + |t - 5\pi|)\sin(2t)$$
$$z = t$$

over the interval $[0, 10\pi]$. Color and remove box and axes.

Solution:

```
//Bed-Spring
clf
t = 0:.01:10*%pi;
a = 5+abs(t-5*%pi);
x = a.*cos(2*t);
y = a.*sin(2*t);
z = t;
param3d(x,y,z,35,75,"X@Y@Z",[4,4])
```

```
e=gce();
e.foreground=color('brown');
e.thickness=3;
title('Bed-Spring')
a=gca();
a.axes_visible="off";
a.box="off";
a.x_label.text =" ";
a.y_label.text =" ";
a.z_label.text =" ";
```

□

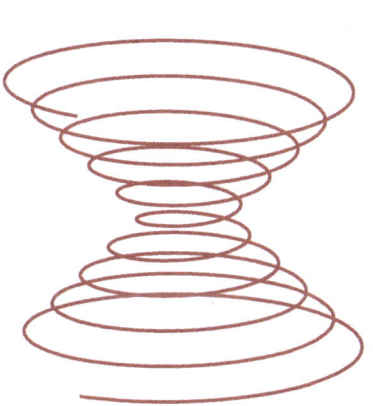

Figure 5.3: Bed-Spring

5.1.4 Draw the toroidal spiral

$$x = (4 + \sin(80t)) \cos(3t)$$
$$y = (4 + \sin(80t)) \sin(3t)$$
$$z = \cos(80t)$$

for $0 \le t \le 4\pi$.

Solution:

```
clf
t =0:%pi/1000:4*%pi;
```

```
x=(4+sin(100*t)).*cos(1*t);
y=(4+sin(100*t)).*sin(1*t);
z=cos(100*t);
param3d(x,y,z/2,30,75,"X@Y@Z",[6,4])
e=gce(); //the handle on the 3D polyline
e.foreground=color('blue');
e.thickness = 2;
```

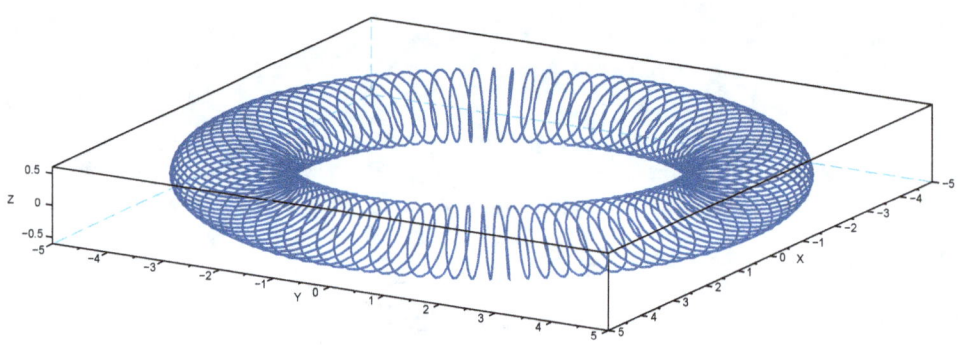

Figure 5.4: Toroid curve

5.1.5 Design an olympic log using the previous toroidal curve.

Solution:

```
//Olympic Logo
clf
t =-0:%pi/1000:4*%pi;
x=(4+sin(100*t)).*cos(1*t);
y=(4+sin(100*t)).*sin(1*t);
z=cos(100*t);
param3d(x,y,z/2,30,75,"X@Y@Z",[6,4])
e=gce(); //the handle on the 3D polyline
e.foreground=color('blue');
e.thickness = 2;

param3d(x+11,y,z/2,270,20)
e=gce();
e.foreground=color('black');
e.thickness = 2;
```

```
param3d(x+22,y,z/2,270,20)
e=gce();
e.foreground=color('red');
e.thickness = 2;

param3d(x+6,y-3,z/2,270,20)
e=gce();
e.foreground=color('yellow');
e.thickness = 2;

param3d(x+17,y-3,z/2,270,20,"X@Y@Z",[6,2])
e=gce();
e.foreground=color('green');
e.thickness = 2;
a=gca;
a.box="off":
```
—

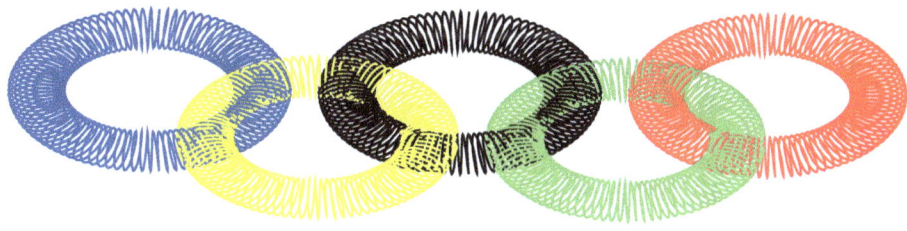

Figure 5.5: Olympic Logo

5.1.6 The Borremean rings are topological knots are given parametrically by:

$x = -a + \cos(t); y = \sin(t); z = b\cos(3t)$; Left ring
$x = a + \cos(t); y = \sin(t); z = b\cos(3t)$; Right ring
$x = \cos(t); y = \sin(t) - \sin(\pi/3); z = b\cos(3t)$; Left ring

where $a = 0.5$, $b = 0.2$, $0 < t < 2\pi$. Graph these rings and **view them from different angles**.

Solution:

```
clf
// Borromean Rings Link
a=0.5; b=0.2;
t=0:.01:2*%pi;
// Left ring
x=-a+cos(t); y= sin(t); z=b* cos( 3*t);
param3d(x,y,z)
e=gce(); e.foreground=color('blue');
e.thickness = 4;
// Right ring
x=a+cos(t); y= sin(t); z=b* cos( 3*t);
param3d(x,y,z)
e=gce(); e.foreground=color('red');
e.thickness = 8;
// Lower ring
x=cos(t); y= sin(t) - sin(%pi/3) ; z=b* cos( 3*t);
param3d(x,y,z,10,75,"X@Y@Z",[6,2])
e=gce(); e.foreground=color('green');
e.thickness = 12;
a=gca(); //get the current axes
a.box="off";
```

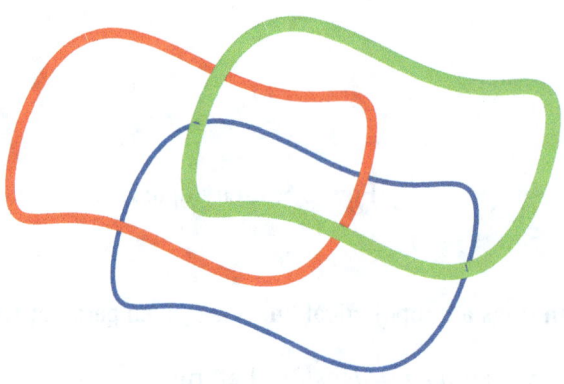

Figure 5.6: Borromean Knots

5.1.7 Graph the Trefoil topological knot which is given parametrically by:

$$x = \cos(pt)(r + \cos(qt))$$
$$y = \sin(pt)(r + \cos(qt))$$
$$z = \sin(qt)$$

where $p = 2$, $\quad q = 3$, $\quad r = 3$, $\quad 0 < t < 2\pi$.

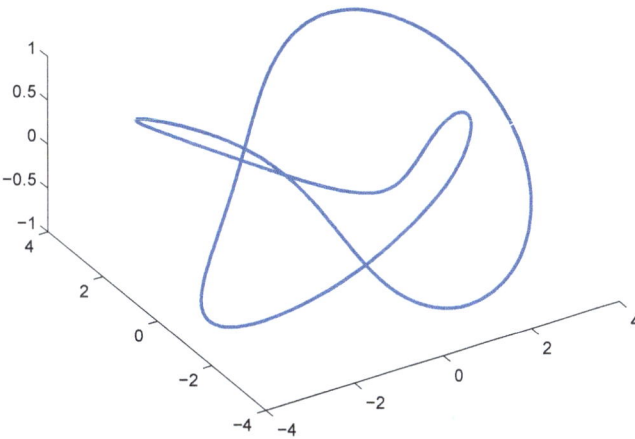

Figure 5.7: Trefoil Knot

5.1.8 The Tornado like curve can be simulated by the equations:

$$r = e^{-(a(t-18))^2};$$
$$x = r\cos(30t) * (1 - a\cos(t)) + (1 - a\cos(t));$$
$$y = r\sin(30t) * (1 - a\sin(t)) + (1 - a\sin(t));$$
$$z = t;$$

where $a = 0.1$, $\quad 0 \leq t \leq 8\pi$. Graph the Tornado curve with different colors.

Solution:

```
// Tornado
clf
t=0:0.001:8*%pi;
r=exp(-0.01*(t-18).^2);
//r=exp(-.2*t);
z=t;
```

```
x=r.*cos(30*t).*(1-.1*cos(t))+(1-.1*cos(t));
y=r.*sin(30*t).*(1-.1*sin(t))+(1-.1*sin(t));
param3d(x,y,z,35, 15, '',[2,0])
e=gce() //the handle on the 3D polyline
//e.foreground=color('red');
e.foreground=color('dimgray');
//e.foreground=color('gray38');
a=gca(); //the handle on the axes
e.thickness = 1.5;
title("Tornado")
```

Tornado

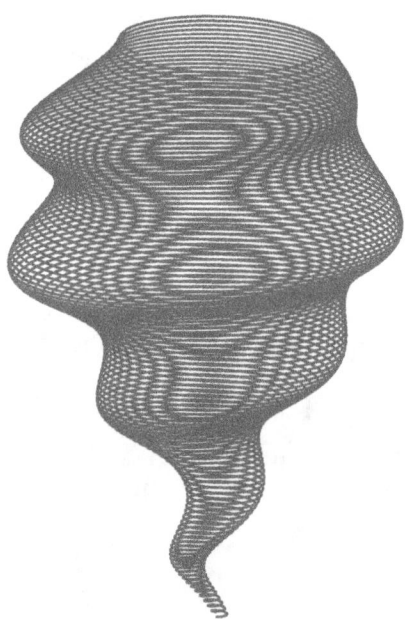

Figure 5.8: Turnado curve

Chapter 6

Surface Plots

Surface plots are used to represent data in a matrix format as a surface. The plots of two-dimensional functions $z = f(x,y)$ are surfaces in three-dimensional space. Scilab has built-in several three-dimensional plots with different options:

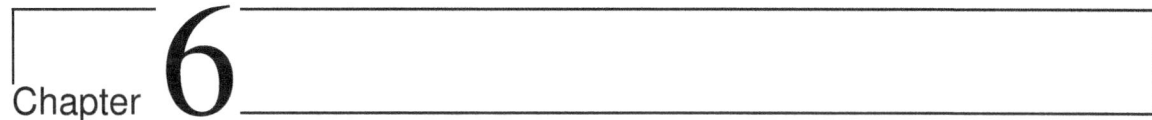

contour(x,y,z,nz, ...)	level curves on a 3D surface.
eval3dp(f,x,y)	evaluates the function f at two vectors x, y and it will output three matrices X, Y, Z.
fplot3d(x,y,f, . . .)	plots surface defined by the function f with two row vectors x, y.
hist3d(X,. . .)	3D representation of a histogram.
mesh(X,Y,Z)	3D mesh plot for matrices X, Y, Z.
plot3d(x, y, z, . . .)	3D plot of a surface for matrix z and two vector rows x, y.
scatter3(x, y, z, . . .)	3D scatter plot for vectors x, y, z.
surf(X,Y,Z, . .)	3D surface plot for matrices Z, X, Y.

6.0.1 Mesh and Surface Plots

Scilab defines a surface by the z-coordinates of points above a grid in the x-y plane, using straight lines to connect adjacent points. The basic two functions to plot surfaces in three dimensions are: *mesh* and *surf* commands.

- *mesh* makes wireframe surfaces that color only the lines connecting the points.
- *surf* displays and colors both the connecting lines and the faces of the surface.

To display a function of two variables, $z = f(x,y)$ with mesh or surf functions, you need to generate X and Y matrices consisting of repeated rows and columns, respectively, over the domain of the function. Scilab has two functions to generate such grids: *meshgrid* and *ndgrid*, the functions transform the domain specified by a single vector or two vectors x and y into matrices X and Y for use in evaluating functions of two variables. The rows of X are copies of the vector x and the columns of Y are copies of the vector y. The function ndgrid is more general than meshgrid. The following example shows the meshgrid at work:

```
x = -2:1:3; // choose x-values
y = 5:1:9;  // choose y-values
[X, Y] = meshgrid(x, y); // constructs matrix format
x , y, X, Y


x  =
 - 2.   - 1.    0.     1.     2.     3.
y  =
   5.     6.     7.     8.     9.
X  =
 - 2.   - 1.    0.     1.     2.     3.
 - 2.   - 1.    0.     1.     2.     3.
 - 2.   - 1.    0.     1.     2.     3.
 - 2.   - 1.    0.     1.     2.     3.
 - 2.   - 1.    0.     1.     2.     3.
Y  =
   5.     5.     5.     5.     5.     5.
   6.     6.     6.     6.     6.     6.
   7.     7.     7.     7.     7.     7.
   8.     8.     8.     8.     8.     8.
   9.     9.     9.     9.     9.     9.
```

In the following examples we will demonstrate the usage of mesh and surf functions. Since meshgrid generates matrices, **it is important to use the element by element operations:** $.*, ./, .\wedge$.

6.0.1 Graph the two dimensional *sinc* function over the domain $[-20, 20] \times [-20, 20]$ using mesh and surf.

Solution:

```
clf
x=[-20:.5:20]; y=x;
[X,Y]=meshgrid(x,y);
    r=sqrt(X.^2+Y.^2+%eps);// Add eps to avoid zero division
    Z=sin(r)./r ;
mesh(X,Y,Z)
// Add title
xtitle("$\huge \dfrac{\sin r}{r}  $")
```

□

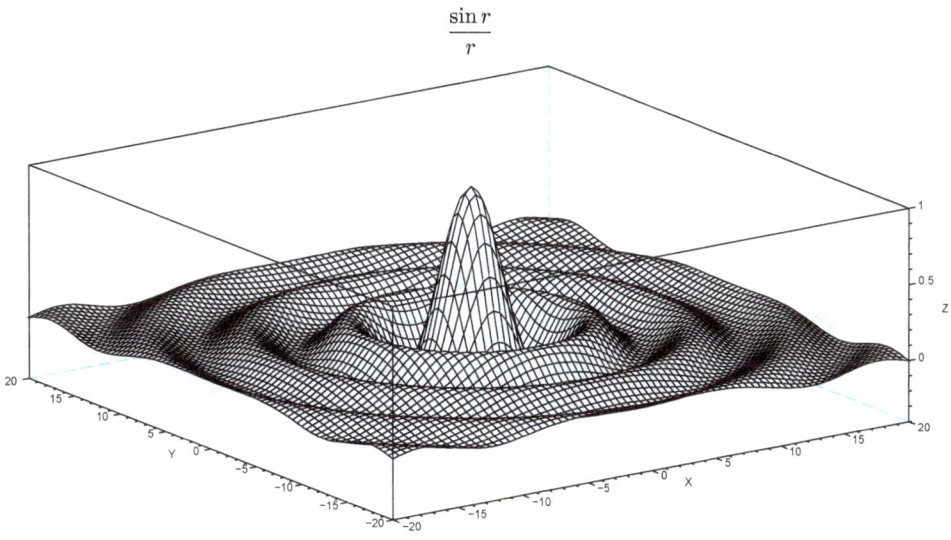

$$\dfrac{\sin r}{r}$$

Figure 6.1: mesh plot

```
clf
surf(X,Y,Z)
xtitle("$\huge \dfrac{\sin r}{r}  $")



// surf plots allow color manipulations
// Add colorbar
clf
Zm = min(Z); ZM = max(Z);
xset("colormap",rainbowcolormap (64))// Check the help for additional colormaps
colorbar(Zm,ZM)
surf(X,Y,Z)
xtitle("$\huge \dfrac{\sin r}{r}  $")
```

6.0.2 plot3d and fplot3d functions

Scilab has two additional surface plots for which the inputs are two vectors x and y and Z matrix is computed directly without grid matrices. In the following example, we graph a surface using mesh, surf and **plot3d**:

6.0.2 Plot the basket surface

$$z = \sqrt{x^8 + y^8}$$

over the domain $-3 <= x <= 3$ and $-3 <= y <= 3$ using mesh, surf, and plot3d commands.

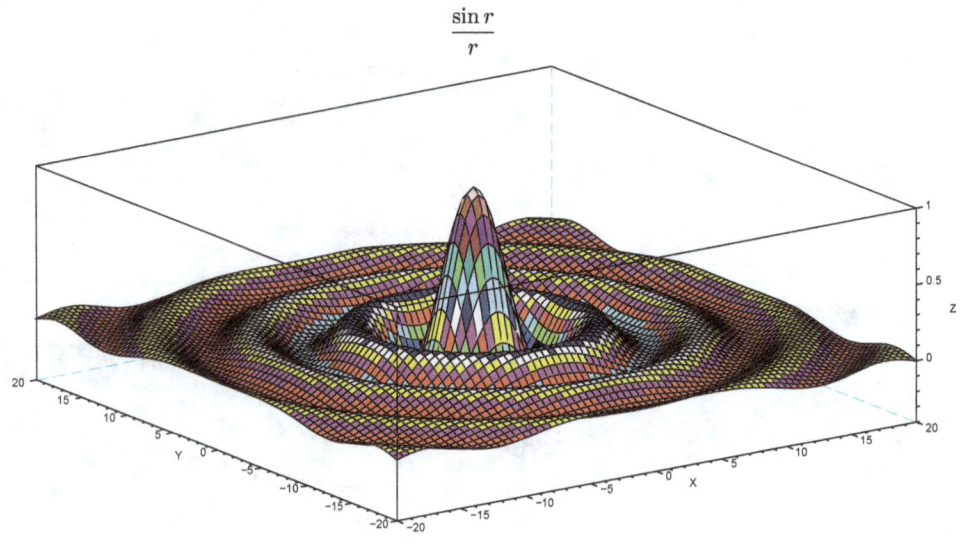

Figure 6.2: surf plot

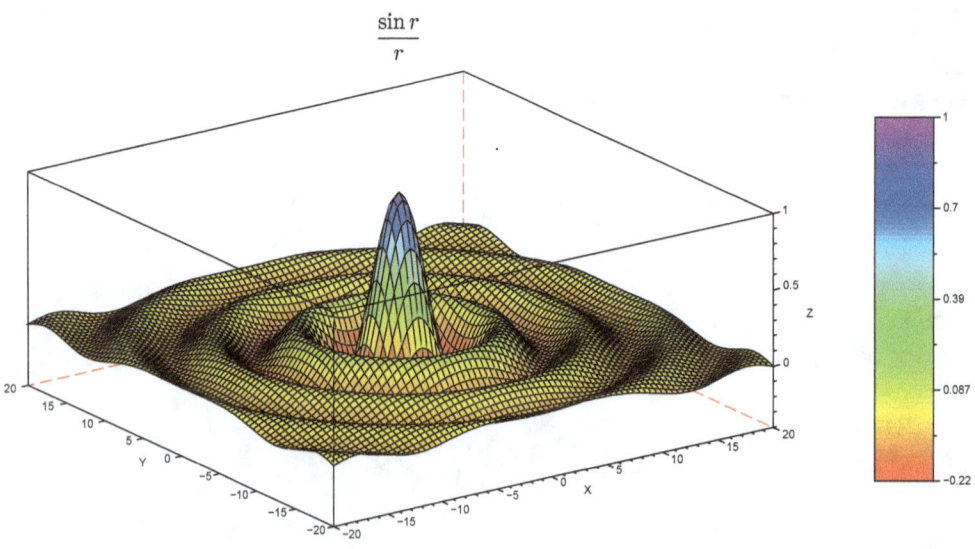

Figure 6.3: colorful surf plot

Solution:

```
clf
x = -3:0.2:3;
y = x;
[X, Y] = meshgrid(x, y);
Z = sqrt(X.^8 + Y.^8)/100 ;//Vector Operations
mesh(X,Y,Z)// matrices X, Y, Z
clf
surf(X,Y,Z)// Matrices X, Y, Z
clf
plot3d(x,y,Z,30,80)// vectors x,y and matrix Z
// Use subplot to combine the three plots
clf
subplot(1,3,1)
mesh(X,Y,Z)// matrices X, Y, Z
subplot(1,3,2)
surf(X,Y,Z)// Matrices X, Y, Z
subplot(1,3,3)
plot3d(x,y,Z,30,80)// vectors x,y and matrix Z
```

$\square$

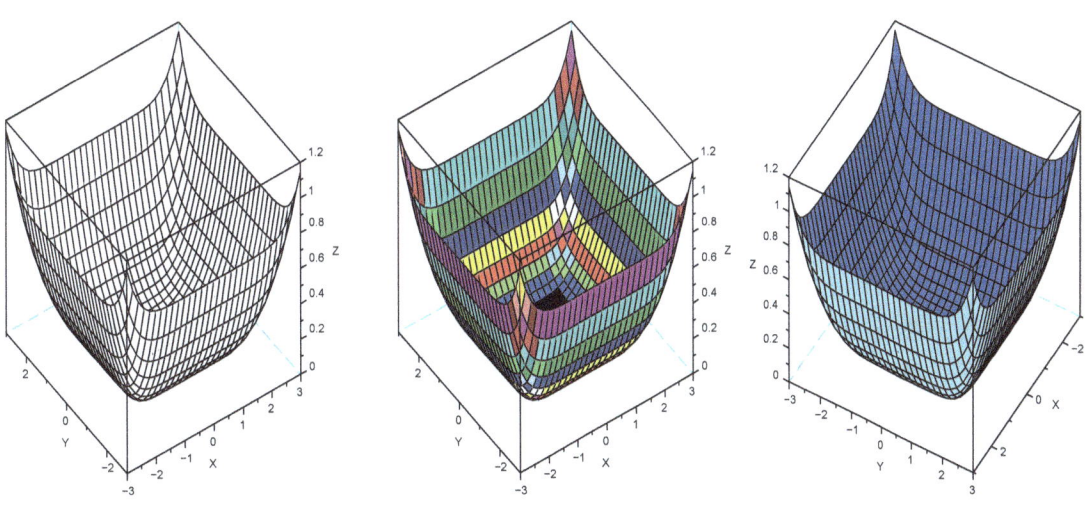

Figure 6.4: mesh, surf and plot3d of basket surface

In the following example, we demonstrate plotting function **fplot3d** applied to the monkey saddle surface.

6.0.3 Graph the surface

$$z = x^3 - 3xy^2$$

known as the monkey saddle surface, using the commands mesh, plot3d, fplot3d, surf , over the domain $[-3,3] \times [-3,3]$. Can you explain why this surface is called monkey saddle?

Solution:

- *mesh, surf requires matrix input using meshgrid or genfac3d commands*
- *fplot3d requires to define the equation as a user-function*

```
//Monkey Saddle Surface
x=linspace(-3,3,30);
y=x;
[X,Y]=meshgrid(x,y);
Z=X.^3 - 3 * X .* Y.^2;
// 1. the mesh command is applied to a matrices X and Y
clf
mesh(X,Y,Z)// Use the pointer to rotate the surface for desired view
// 2. We can use the surf function with matrices inputs X and Y
clf
surf(X,Y,Z)
// 3.  plot3d is used with domain inputs are the vectors x and y
clf
plot3d(x,y,Z)

// 4. fplot3d requires the surface to be defined as a function and the
domain inputs are x and y vectors

function z=fm(x,y)
    z=x^3-3*x*y^2
endfunction

clf
fplot3d(x,y,fm)
// Another variation of fplot3d

clf
fplot3d1(x,y,fm)

// We combine these plots in one figure
clf
subplot(2,2,1)
mesh(X,Y,Z)
xtitle("mesh plot")
subplot(2,2,2)
surf(X,Y,Z,'facecol','red','edgecol','yellow')
xtitle("surf plot with color options")
subplot(2,2,3)
plot3d(x,y,Z)
```

```
xtitle("plot3d")
subplot(2,2,4)
fplot3d1(x,y,fm)
xtitle("fplot3d")
```

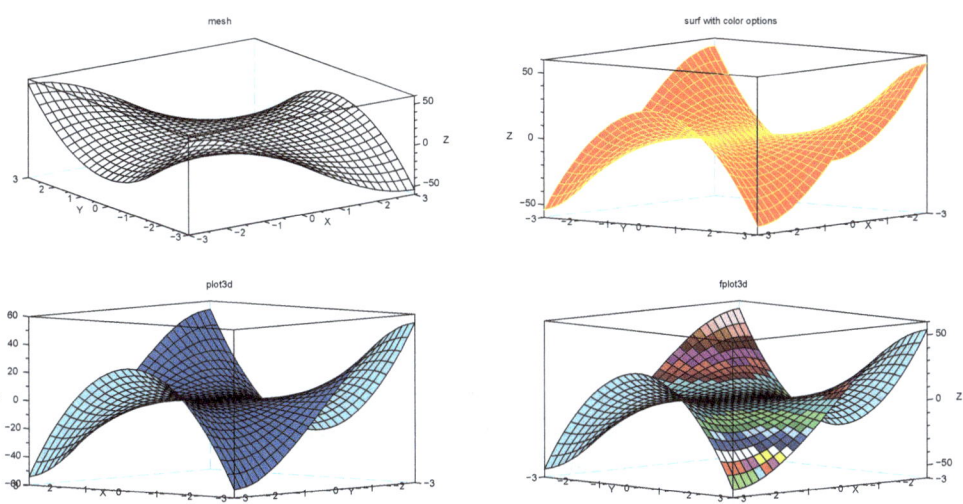

Figure 6.5: Monkey Saddle plots

6.0.4 For all subsequent plots, we will call a user defined function sweep() which helps clear axes and select different color maps:

```
function sweep();
f=gcf();//get the handle of the parent figure
f.color_map=coppercolormap(64);
//f.color_map=rainbowcolormap(32);
h=gce(); //get handle on current entity (here the surface)
h.color_mode=-1;
a=gca(); //get current axes
a.axes_visible="off"; //axes are hidden
a.box="off";
a.x_label.text =" ";
a.y_label.text =" ";
a.z_label.text =" ";
/* list of color maps
autumncolormap: red through orange to yellow colormap
bonecolormap: gray colormap with a light blue tone
coppercolormap: black to a light copper tone colormap
graycolormap: linear gray colormap
hotcolormap: red to yellow colormap
hsvcolormap:Hue-saturation-value colormap
jetcolormap: blue to red colormap
oceancolormap: linear blue colormap
parulacolormap: blue to yellow colormap
pinkcolormap: sepia tone colorization on black and white images
rainbowcolormap: red through orange, yellow, green, blue to violet colormap
springcolormap: magenta to yellow colormap
summercolormap:green to yellow colormap
*/
endfunction
```

6.0.5 Use surf function to view the Horse Saddle

$$z = \frac{(1.2 - x^2)(1.2 - y^3)}{(1 + x^2 + y^2)}$$

over the domain $[-1.5, 2] \times [-1.5, 2]$

Solution:

```
//Horse Saddle Surface
clf
x=[-1.5:%pi/60:2];y=x;
[X,Y]=meshgrid(x,y);
Z=(1.2-X.^2).*(1.2-Y.^3)./(1+X.^2+Y.^2);//Horse Saddle
surf(X,Y,Z)
sweep()
xtitle('Horse Saddle')
```

Horse Saddle

Figure 6.6: Horse Saddle surface

☐

6.0.6 Explore the following canopy-like surface and its intersection with planes. Use different planes and different colors.

Solution:

```
//Intersection of plane with Canopy
//canopy
x=-.65:.1:.67;
y=x;
[X,Y]=meshgrid(x,y);
Z=(1- X.^2 - Y.^2).^(-0.5);
ZP=X+Y-1.5;
clf
surf(X,Y,-Z)
surf(X,Y,ZP)
sweep()
```

☐

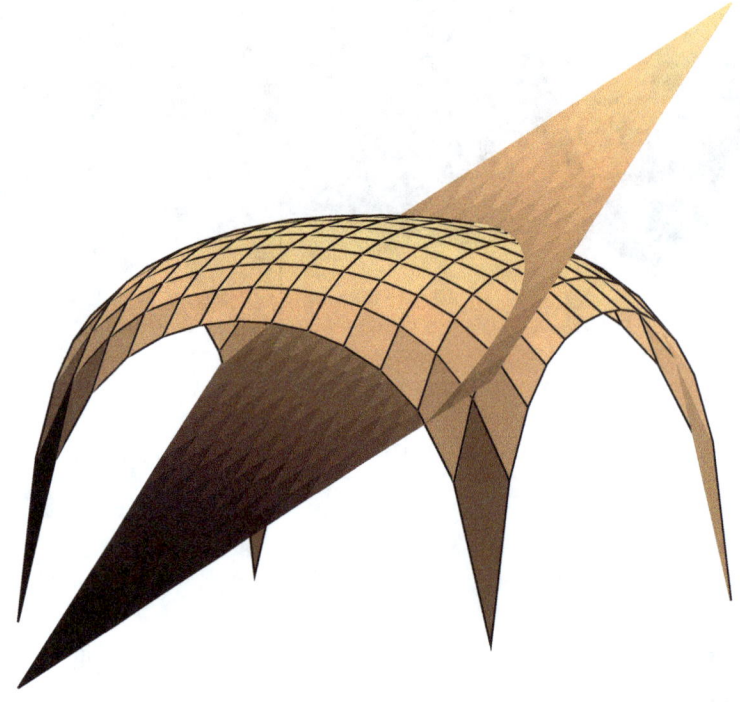

Figure 6.7: Canopy and plane

6.0.3 Contour Plots

Contour plots are another variant of visualizing matrices and three-dimensional surfaces. Contour plots are lines (level curves) which represent constant data or heights. Scilab provides a contour plot function with the option to specify the number of cuts (levels) and various options 2-d or 3-d contours with or without printed levels and surrounding box. In the following example, we will show how to produce contour plots with options.

6.0.7 For the surface

$$z = (x^2 + 2y^2 - 1)e^{1-(x^2+y^2)}$$

over the square $[-3,3] \times [-3,3]$. Produce the following plots:
- Plot the surface with fplot3d1
- Plot 2-d contour plot with 15 levels
- Plot the surface using plot3d and add 3-d contour plot to the surface

- plot 20 levels of 3-d contour plot

Solution:

```
x=-3:.2:3; y=x;
function z=f(x,y),z=exp(-(x^2+y^2)+1)*(x^2+2*y^2 -1.0);endfunction
clf
subplot(2,2,1)
fplot3d1(x,y,f,20,75);
#clf
subplot(2,2,2)
xset("fpf"," ");//To suppress printing levels
contour(x,y,f,15);// 15 levels
isoview(-3,3,-3,3);// square scaling

z=feval(x,y,f);// z matrix
subplot(2,2,3)
plot3d(x,y,z);
xset("fpf"," %.1f");//Print one decimal levels
contour(x,y,z,10,flag=[0 2 4]);// 10 levels

subplot(2,2,4)
xset("fpf"," ");// Omit numerical leveling
contour(x,y,z,20,flag=[0 2 2]);//20 levels
```

☐

6.0.4 Color Shading

Surface plot has options to color the facets and the edges of each rectangular facets. For example,

surf(X,Y,Z, ...)	draws surface with colored faces and edges.
Sgrayplot(x,y,Z)	draws smooth 2d plot of the surface (heat map).

In the following example, we demonstrate some of the color options available in Scilab.

6.0.8 Draw the surface

$$z = \ln(2 + \sin(x^2 - y^2))$$

over the square $[-2,2] \times [-2,2]$.
- Plot the surface with surf command
- Plot the surface with cyan faces and magenta edges levels
- Plot the surface using circle marks in yellow with red edges and blue faces (Try different colors).
- Plot a heat map of the surface and with barcolor.

Solution:

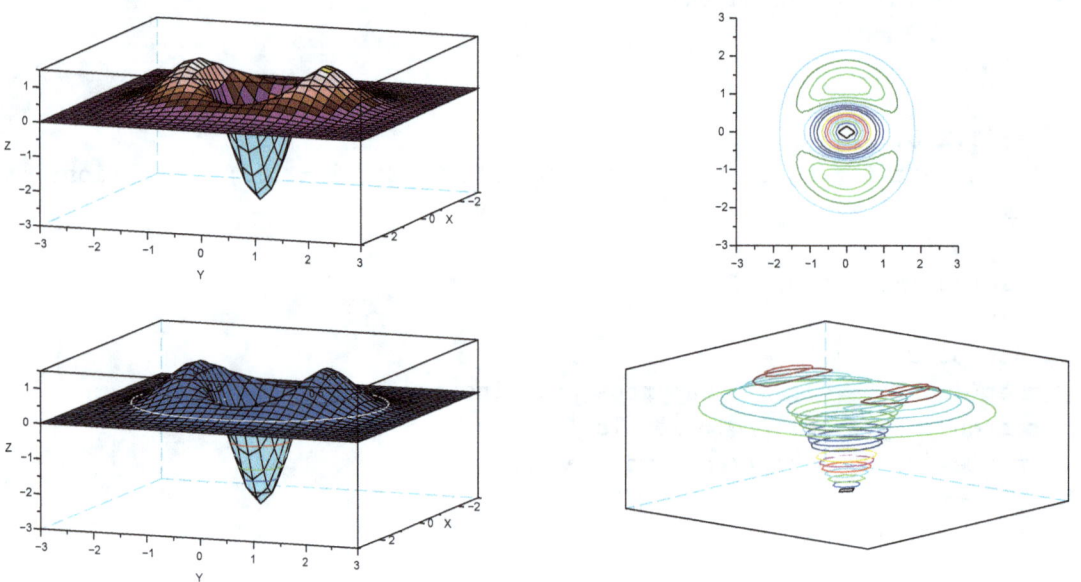

Figure 6.8: Contour Plots

```
x=-2:.1:2;
y=x;
[X,Y]=meshgrid(x,y);
Z=log(2+sin(X.^2-Y.^2));
clf
surf(X,Y,Z) //default
xtitle("$\huge z = \ln(2 + \sin(x^2 - y^2))$")

clf
surf(X,Y,Z,'facecol','cyan','edgecol',
'magenta')
xtitle("$\huge z = \ln(2 + \sin(x^2 - y^2))$")

clf
surf(X,Y,Z,'edgecol','yellow','marker',
'o','markersize',

8,'markeredge','red','facecol','blue')

xtitle("$\huge z = \ln(2 + \sin(x^2 - y^2))$")

clf
//Heat Map
Sgrayplot(x,y,Z)
```

```
clf
//Heat Map with color options
xset("colormap",springcolormap(64))
colorbar(-1,1)
Sgrayplot(x,y,Z, strf="041")
```
□

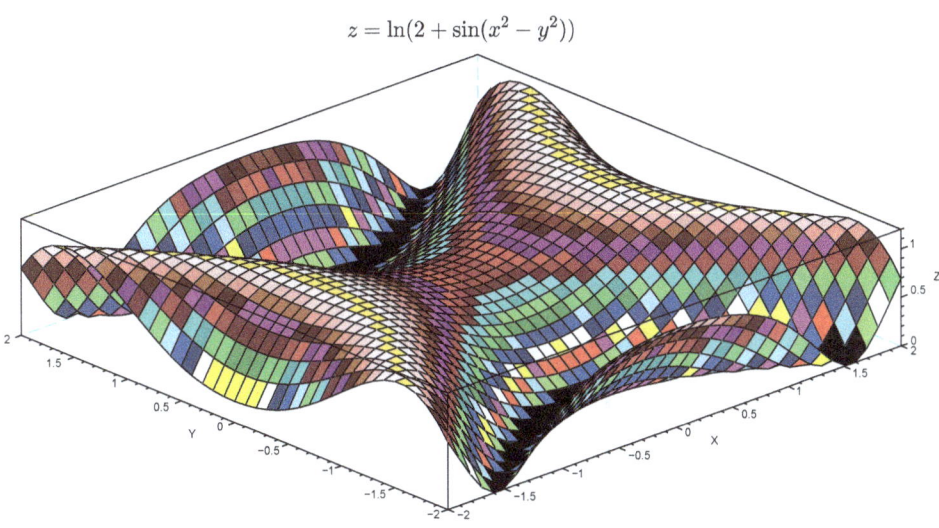

$$z = \ln(2 + \sin(x^2 - y^2))$$

Figure 6.9: Default surface plot

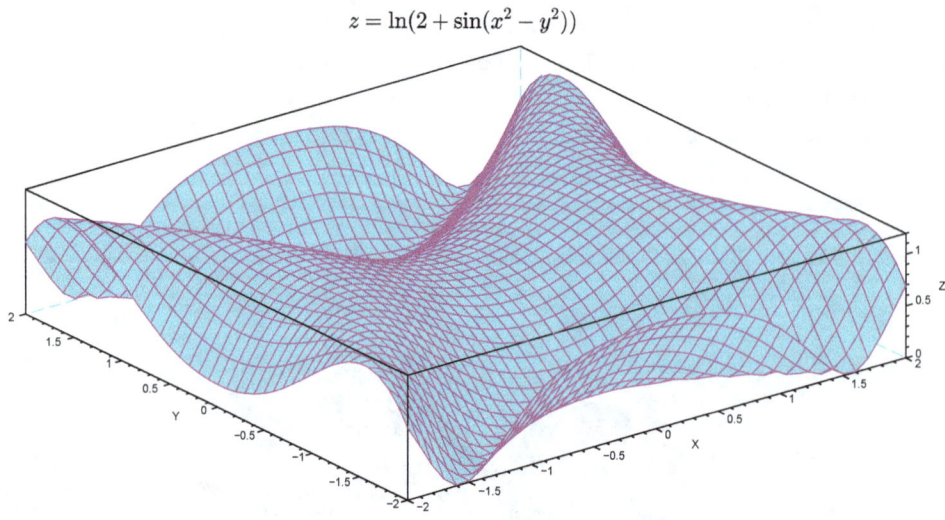

Figure 6.10: Surface plot with coloring faces and edges

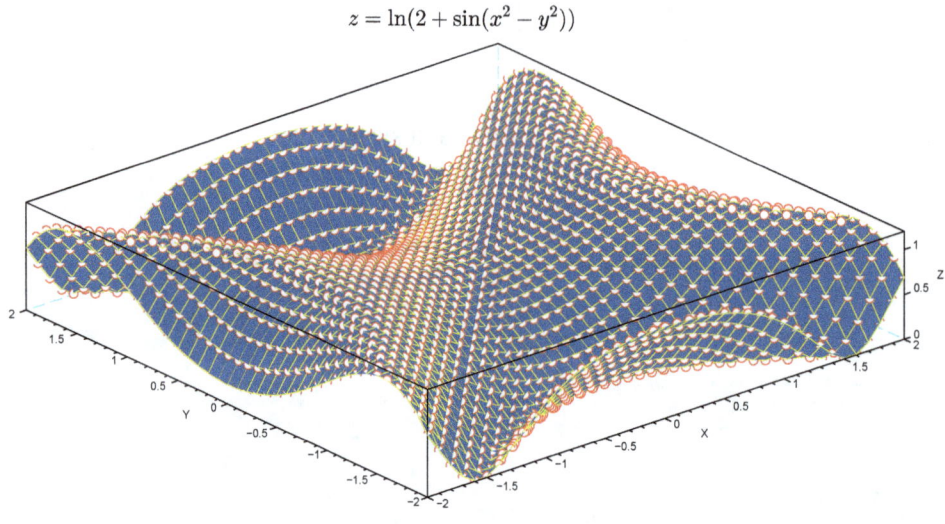

Figure 6.11: Surface plot with colored markers, faces and edges

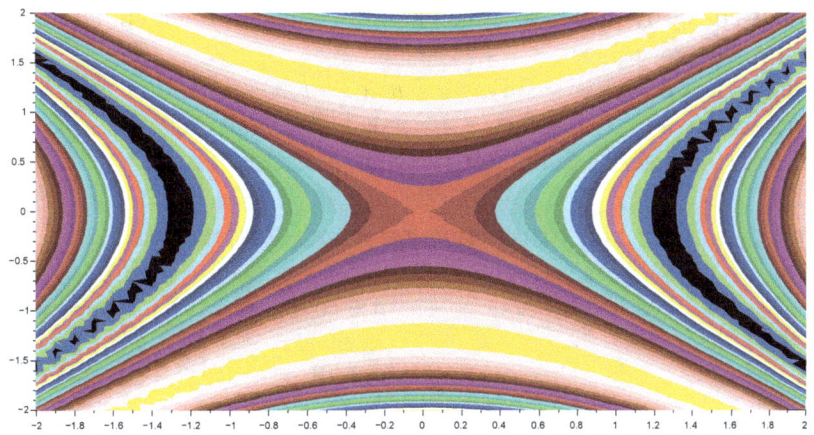

Figure 6.12: Heat Map of the surface

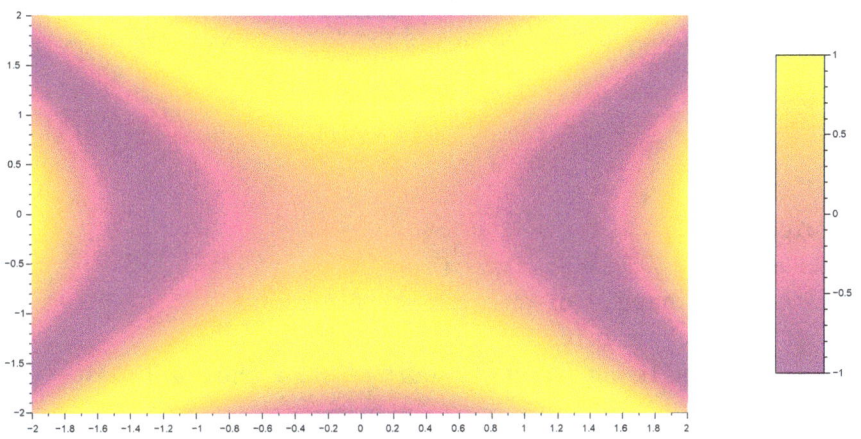

Figure 6.13: Heat Map of the surface with smoothing colors

6.1 Surfaces in parametric form

Parametrized surfaces are easily plotted in Scilab as will be demonstrated in the examples. Here as few examples of parametrized surfaces:

1. Circular cylinder: $x = \cos v$, $y = \sin v$, $z = u$ on $[-\pi, \pi] \times [-\pi, \pi]$
2. Circular cone: $x = u \cos v$, $y = u \sin v$, $z = u$ on $[-\pi, \pi] \times [-\pi, \pi]$
3. Circular helix: $x = u \cos v$, $y = u \sin v$, $z = v$ on $[-\pi, \pi] \times [-\pi, \pi]$
4. Twisted cone:
   $x = (u - \sin u) \cos v$, $y = (1 - \cos u) \sin v$, $z = u$ on $[-\pi, \pi] \times [-\pi, \pi]$
5. Twisted horn:
   $x = (1 - u)(3 + \cos v) \cos(4\pi u)$, $y = (1 - u)(3 + \cos v) \sin(4\pi u)$, $z = 3u + (1 - u) \sin v$ on $[-1, 1] \times [-1, 1]$
6. Twisted cylinder:
   $x = (2 + \sin v) \cos u$, $y = (2 + \sin v) \sin u$, $z = u + \cos v$ on $[-\pi, \pi] \times [-\pi, \pi]$
7. Miobus strip:
   $x = 2 \cos v + u \cos(v/2)$, $y = 2 \sin v + u \cos(v/2)$, $z = u \sin(v/2)$ on $[-.5, .5] \times [0, 2\pi]$
8. Dini's surface:
   $x = \cos u \sin v$, $y = \sin u \sin v$, $z = \cos v + \log(\tan(v/2)) + u/(2\pi)$ on $[2\pi, 6\pi] \times [0, \pi/2]$
9. Spheres:
   $x = a \sin u \cos v$, $y = a \sin u \sin v$, $z = a \cos v$ on $[0, \pi] \times [0, 2b\pi]$

6.1.1 Three pipe calculus Problem: The coordinate axes of three right circular cylinders with radius one. Find the volume of the solid that is common to the three cylinders. Here, just visualize these pipes.

Solution:

```
//Three pipes
function sweep(); //cleaner function
f=gcf();//get the handle of the parent figure
f.color_map=coolcolormap(64);
h=gce(); //get handle on current entity (here the surface)
h.color_mode=-1;
a=gca(); //get current axes
a.axes_visible="off"; //axes are hidden
a.box="off";
a.x_label.text =" ";
a.y_label.text =" ";
a.z_label.text =" ";
endfunction

st=.1;
function [x,y,z]=f(u,v);
x=cos(v);
y=sin(v);
z=u;
endfunction
```

```
u1=-%pi:st:%pi+st;v1=u1;
[x1,y1,z1]=eval3dp(f,u1,v1);
clf
plot3d1(x1,y1,z1);
sweep()

plot3d1(2*z1,x1,y1);
sweep()

plot3d1(x1,1.5*z1,y1)
sweep()
xtitle('Three Pipes ')
```

□

Figure 6.14: Three pipes

6.1.2 Twisted Surfaces. Visualizations of a circular cone, and twisted helix, cone, and cylinder.

Solution:

```
//Axes and Color functions
function noaxes();
a=gca(); //get current axes
a.axes_visible="off"; //axes are hidden
a.box="off";
a.x_label.text =" ";
a.y_label.text =" ";
a.z_label.text =" ";
endfunction
//
function sweeper();
f=gcf();//get the handle of the parent figure
f.color_map=rainbowcolormap(64);
h=gce(); //get handle on current entity (here the surface)
h.color_mode=-1;
noaxes()//noaxes() is defined before
endfunction

// circularity cone
clf
subplot(2,2,1)
st=.1;
function [x,y,z]=f2(u,v);
    x=u.*cos(v);
    y=u.*sin(v);
    z=u;
endfunction
u=-%pi:st:%pi+st;v=u;
[x,y,z]=eval3dp(f2,u,v);
plot3d1(x,y,z);
noaxes()
xtitle('Circular cone')
// circular helix
subplot(2,2,2)

function [x,y,z]=f2(u,v);
    x=u.*cos(v);
    y=u.*sin(v);
    z=v;
endfunction
u=-%pi:st:%pi+st;v=u;
[x,y,z]=eval3dp(f2,u,v);
plot3d(x,y,z)
noaxes()
xtitle('Circular helix ')
// twisted cone
```

```
subplot(2,2,3)
function [x,y,z]=f2(u,v);
     x=(u-sin(u)).*cos(v);
     y=(1-cos(u)).*sin(v);
     z=u;
endfunction
u=-%pi:st:%pi+st;v=u;
[x,y,z]=eval3dp(f2,u,v);
plot3d1(x,y,z);
sweeper()
xtitle('Twisted cone ')
//  twisted cylinder
subplot(2,2,4)
function [x,y,z]=f2(u,v);
x=(2+sin(v)).*cos(u);
y=(2+sin(v)).*sin(u);
z=u+cos(v);
endfunction
u=-%pi:st:3*%pi+st;v=-%pi:st:%pi+st;
[x,y,z]=eval3dp(f2,u,v);
plot3d1(x,y,z);
sweeper()
xtitle('Twisted cylinder   ')
```

□

6.1.3 Topological Surfaces. Draw the torus as an example of an orientable surface, followed by the Mobius strip as an example of a nonorientable surface with boundary: it has one side and is physically constructed by hand by half twisting and taping together the ends of a piece of paper. also, include a twisted horn surface.

Solution:

```
//Use sweeper() and noaxes() from previous example
clf
// Twisted horn
subplot(3,1,1)
function [x,y,z]=f2(u,v);
x=(3+u.*cos(v)).*sin(2*%pi*u);
y=(3+u.*cos(v)).*cos(2*%pi*u)+2*u;
z=u.*sin(v);
endfunction

u=linspace(0,1,80);
v=linspace(0,2*%pi,80);
[x,y,z]=eval3dp(f2,u,v);
plot3d1(x,y,z);
```

Circular cone

Circular helix

Twisted cone

Twisted cylinder

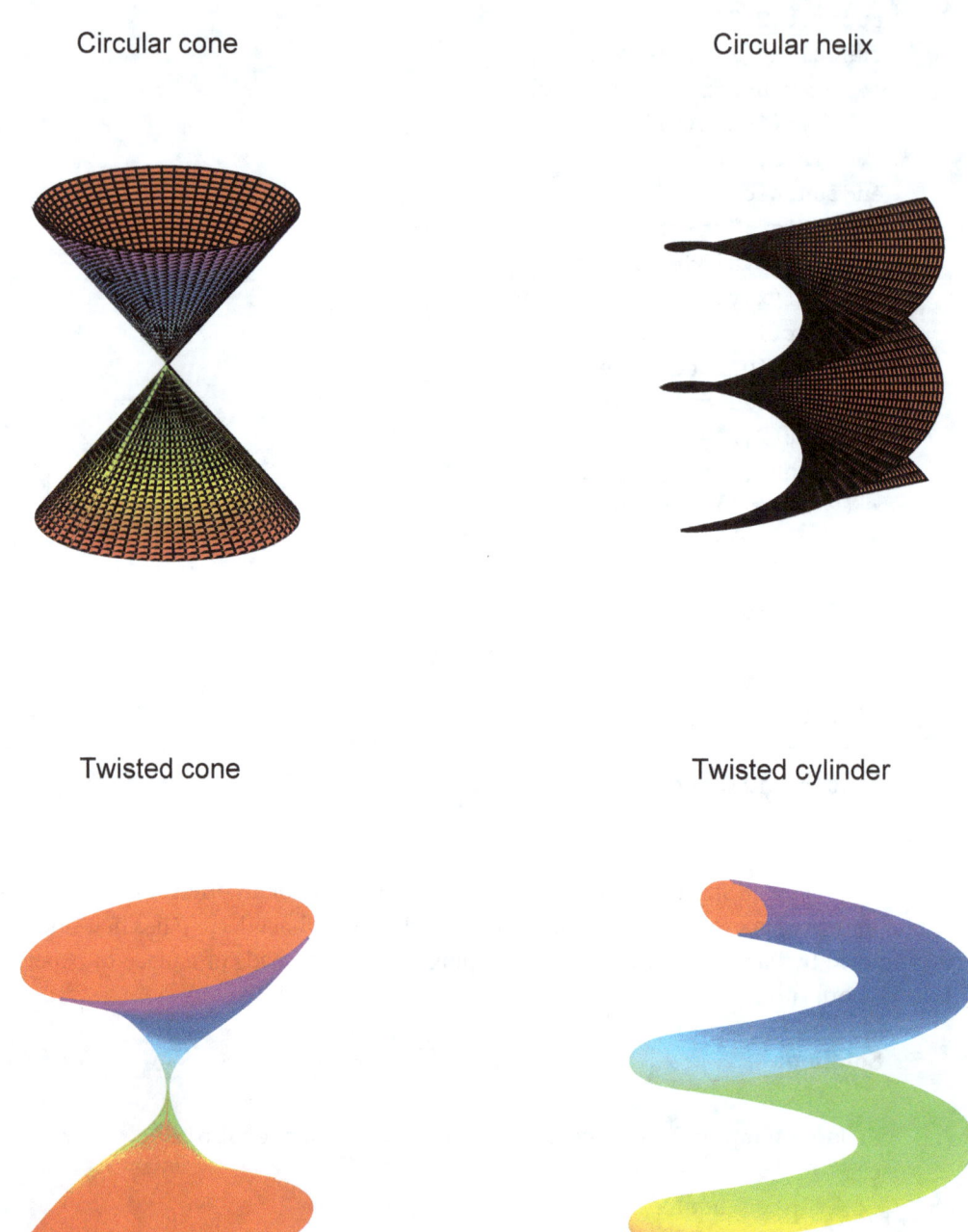

Figure 6.15: Twisted surfaces

```
sweeper()
//plot3d1(x1,12-y1,z1);
xtitle('Twisted horn  ')
```

```
// Torus
subplot(3,1,2)
st=.05;
function [x,y,z]=f2(u,v);
    a=3;b=1;
    x=(a+b*cos(v)).*cos(u);
    y=(a+b*cos(v)).*sin(u);
    z = a*sin(v);
endfunction
u=linspace(-%pi,%pi,50);
v=u;
[x,y,z]=eval3dp(f2,u,v);
surf(x,y,z,'facecol','interp')
noaxes()
xtitle('Torus    ')

// Mobius strip
subplot(3,1,3)
st=.05;
function [x,y,z]=f2(u,v);
    x=2*cos(v)+u.*cos(v/2);
    y=2*sin(v)+u.*cos(v/2);
    z=u.*sin(v/2);
endfunction
u=-.5:st:.5;v=0:st:2*%pi;
[x,y,z]=eval3dp(f2,u,v);
plot3d1(x,y,z);
sweeper()
xtitle('Miobus strip    ')
```

□

Twisted horn

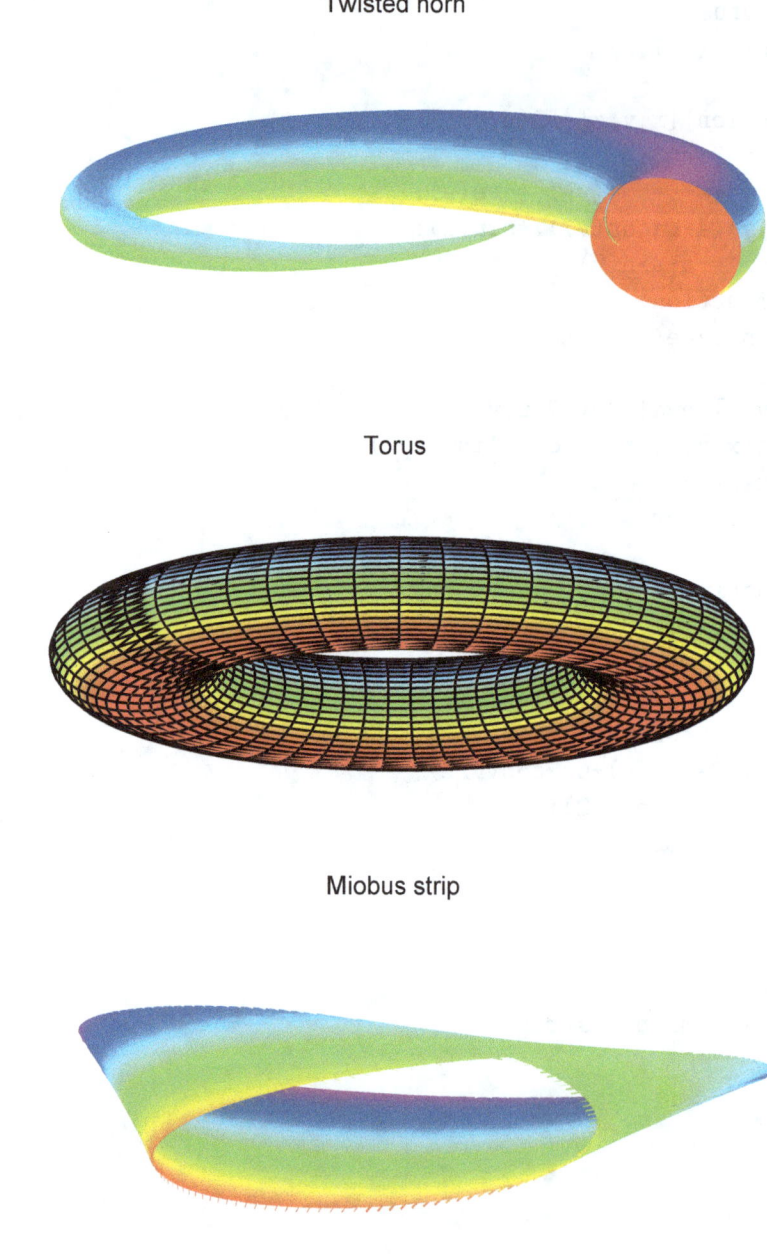

Torus

Miobus strip

Figure 6.16: Topological surfaces

6.2 Solids of Revolution

Surfaces of revolutions are generated by revolving a function $y = f(x)$ about the $x - axis$ over an interval $[a, b]$. Such surface is described parametrically by

$$x = u, \qquad \qquad (6.1)$$

$$y = f(u)\cos(v), \qquad \qquad (6.2)$$

$$z = f(u)\sin(v) \qquad \qquad (6.3)$$

where $a \le u \le b$ and $0 \le v \le 2\pi$

6.2.1 Solids of revolution. Plot the function $y = x\sin^2(x)$ over $[0, 2.6]$ and plot the surface generated by revolving this function about the x-axis and the y-axis.

Solution:

```
//Use sweeper() and noaxes() from previous example
clf
//Cylindrical surfaces
st=.1;
function y=g(x)
    y=x.*sin(x);
endfunction

clf
subplot(3,1,1)
x=0:.01:%pi-0.4;
plot(x,-g(x))
function [x,y,z]=f2(u,v);
    x=(u).*cos(v);
    y=(u).*sin(v);
    z=g(u);
endfunction

u= 0:st:5*%pi/6;
v=-%pi:st:%pi+st;
[x,y,z]=eval3dp(f2,u,v);

subplot(3,1,2)
plot3d(x,y,-z)
noaxes()

function [x,y,z]=f2(u,v);
    y=g(u).*cos(v);
    z=g(u).*sin(v);
    x=u;
endfunction


u=0:st:%pi;
```

```
v=-%pi:st:%pi+st;
[x,y,z]=eval3dp(f2,u,v);

subplot(3,1,3)
plot3d(x,y,z)
noaxes()
```

☐

6.2.2 Gabriel's Horn. Draw the solid of revolution generated y the function $y = 1/x, [1, \infty]$ about x-axis. In calculus, we can show that the surface area of Gabriel's Horn is infinite but that its volume is finite.

Solution:

```
//Use sweeper() and noaxes() from previous example
clf
//Gabriel Horn
st=.1;
function y=g(x)
    y=1./x;
endfunction
function [x,y,z]=f2(u,v);
    x=u;
    y=g(u).*cos(v);
    z=g(u).*sin(v);
endfunction

u= 1:st:10;
v=-%pi:st:%pi+st;
[x,y,z]=eval3dp(f2,u,v);
clf
plot3d(x,y,z)
noaxes()
```

☐

6.3 Homework: 3d Plots

6.3.1 Graph the parametric curve

$$x = \sin(t), \qquad y = \sin(2t), \qquad z = \sin(6t)$$

for $-2\pi \leq t \leq 2\pi$ with step size $.02\pi$ and graph it.

6.3.2 Graph the parametric curve

$$x = (1 + \cos 16t) \cos t, \qquad y = (1 + \cos 16t) \sin t, \qquad z = (1 + \cos 16t)$$

for $-2\pi \leq t \leq 2\pi$ with step size $.02\pi$ and graph it.

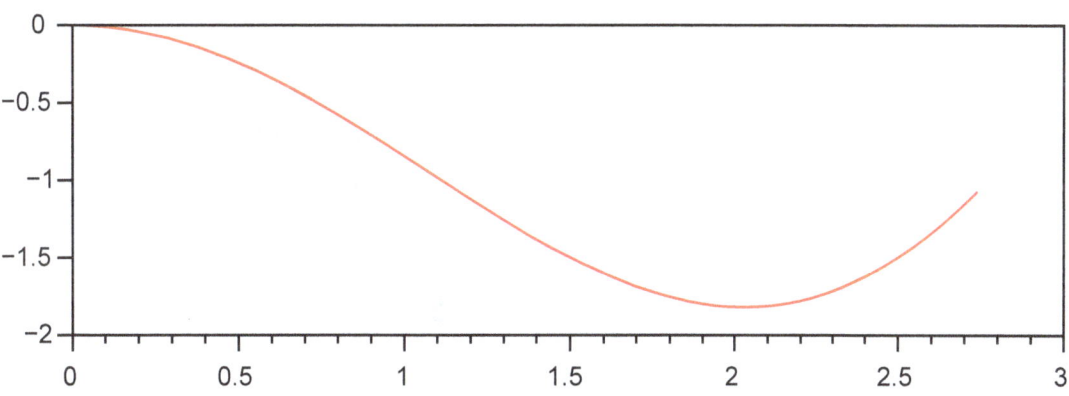

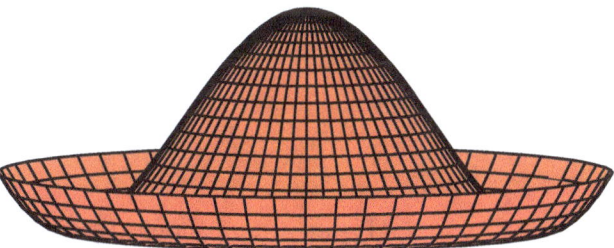

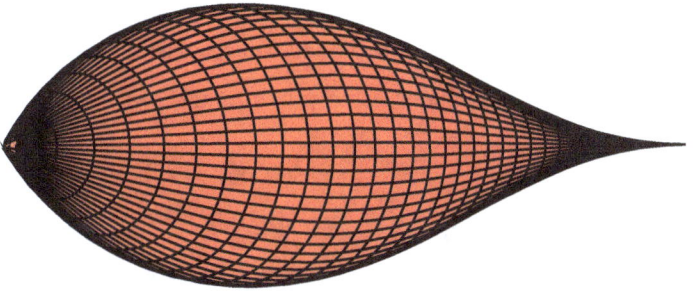

Figure 6.17: Surfaces of revolution

6.3.3 Graph the surface by using fplot3d and rotate it

$$z = -xye^{-(x^2+y^2)}$$

on the square $[-3,3] \times [-3,3]$. Draw the level curves using contour command.

Figure 6.18: Gabriel's Horn

6.3.4 Graph the dog saddle using fplot3d and rotate it

$$z = xy^3 - yx^3$$

on the square $[-3, 3] \times [-3, 3]$. Draw the level curves using contour command.

6.3.5 Graph the circular cylinder using plot3d and replot to generate three intersecting pipes with different colors

$$x = \cos v; y = \sin v; z = u;$$

over the square $[-\pi, \pi] \times [-\pi, \pi]$ with step size 0.3.

6.3.6 Plot the vector field $(y, -x)$ using champ1 function on the square $[-2, 2] \times [-2, 2]$.

6.3.7 Use contour to do the following:
1. Plot the level curves of the function $f(x, y) = 2x^2 - 2y^2 - 3x^3 + 3y^3$ in the square between -1 and 1. Describe the behavior o the surface near zero and in the larger region
2. Plot the level curves of the function $f(x, y) = x \ln y - y \ln x$ that contains the point (e, e)

6.3.8 Plot the surfaces :
1. $z = \cos x \cos y$, for $-5\pi \leq x \leq 5\pi$, $-5\pi \leq y \leq 5\pi$
2. $z = (x^2 + y^2) \sin(x^2 + y^2)$, for $-2 \leq x \leq 2$, $-2 \leq y \leq 2$

6.3.9 Plot the three dimensional curve $x = e^{-t} \cos(t), y = e^{-t} \sin(t), z = t - 10$ over $0 \leq t \leq 20$.

6.3.10 Plot the three dimensional curve

$$x = \cos t, \qquad y = \sin t, \qquad z = \sin 2t$$

for $0\pi \le t \le 2\pi$ with a spacing of $.01\pi$.

6.3.11 Let

$$f(x,y) = \frac{x^2 - 2y^2}{x^4 + 2y^2}$$

Graph the surface $f(x,y)$

6.3.12 A parametrization of Umbilic Torus is given by

$$x = [7 + \cos(\tfrac{1}{3}s - 2t) + 2\cos(\tfrac{1}{3}s + 2t)]\sin s$$
$$y = [7 + \cos(\tfrac{1}{3}s - 2t) + 2\cos(\tfrac{1}{3}s + 2t)]\cos s$$
$$z = \sin(\tfrac{1}{3}s - 2t) + 2\sin(\tfrac{1}{3}s + 2t)$$

Graph this torus on the square $-\pi \le s \le \pi$, $-\pi \le t \le \pi$.

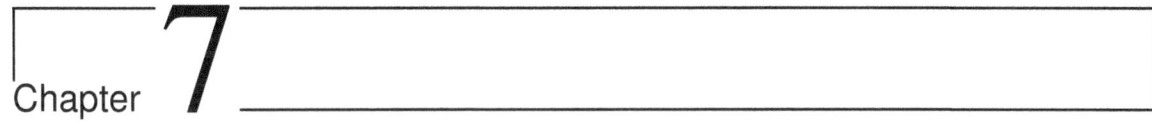

Chapter 7

Data Analysis and Statistics

In this chapter, we present the main Scilab commands for data manipulation and statistical analysis of data sets.

The Scilab Statistics module consists of many built-in various statistical functions grouped in the sections:

- Cumulative Distribution Functions
- Central Tendency Measures
- Descriptive Statistics
- Measures of Dispersion and Shape
- Linear Regression, Sampling, and Testing
- Data Visualization: histogram, bar, pie, and scatter graphs
- Random Numbers Generators

7.1 Data Entry

Data sets can be typed by the user using SciNotes, other text editors, spreadsheet or they can be simulated. However, most data are imported from other sources and in various formats. Data will be assumed in rectangular table format. We suggest creating a new folder to contain your data sets and making it the current working directory, so Scilab will access the data without specifying the path. The basic Scilab commands to handle data sets:

(M) = read("filename",m,n)	reads data in matrix format, m,n are the number of rows and columns to be read, use $m = -1$ to read all rows.
write("filename",A)	writes the matrix A to a filename.
save("filename",x1,x2,...,xn)	saves the Scilab current variables $x_1, x_2, \cdots x_n$ in a binary file.
load("filename" x1,...,xn)	reload in the Scilab session variables previously saved in a file with the save command.
M = csvRead("filename", sep)	read comma-separated value file. It has other options such as header:M = csvRead(filename, separator, decimal, conversion, substitute, regexp comments, range, header).

7.1.1 User typed data We will collect the actual temperature of three cities: New York, Moscow and Cairo from June 1, 2019 to June 15, 2019 and combine them into a data set and save it by the name temps and reload it.

Solution: *We type the entries by hand in a SciNotes editor:*

```
NewYork = [27 27 21 21 27 28 27 27 26 21 25 23 18 23 27]';
Moscow = [23 26 22 24 27 29 31 31 31 27 28 24 20 21 21]';
Cairo = [37 37 37 34 33 35 33 33 34 36 39 33 33 32 35]';
Temps = [NewYork, Moscow, Cairo]
write ("Temps3", Temps)
clear
Temps
Temps=read("Temps3",-1,3);
size(Temps)
plot(Temps)
legends([' Cairo';'New York';'Moscow'],[5,3,2],opt="ur")
xlabel("1-15 June, 2019", "fontsize", 2);
ylabel("Temperature", "fontsize", 2);
```

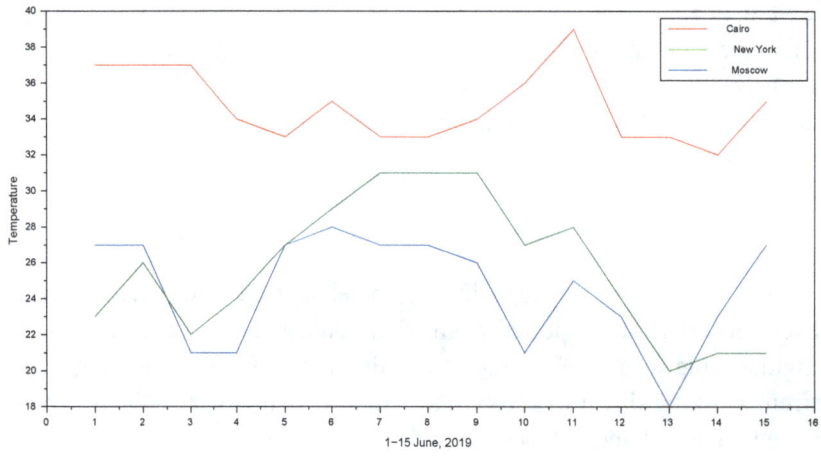

Figure 7.1: Three Cities Temperature

7.1.2 Loading csv files Copy the matrix Temps from last example to a note pad editor and save it to a text file *Temps.txt*, then copy it to an excel sheet and save it in the format *CSV(Comma delimited)(*.csv): Temps.csv*. Load both files to Mtext matrix and Mcsv matrix and check print the first 3 rows.

Solution: *After using the editors to copy and paste the temperature matrix to t files: Temps.txt and Temps.csv in the data working directory*

```
--> Mtext =read('temps.txt',-1,3);
```

```
--> Mtext(1:3,:)// first three rows
  ans  =
    27.   23.   37.
    27.   26.   37.
    21.   22.   37.
--> // reading csv file has few options, "," separator
--> Mcsv = csvRead("Temps.csv",",");
--> Mcsv(1:3,:)
  ans  =
    27.   23.   37.
    27.   26.   37.
    21.   22.   37.
```

□

Web Data

The internet has expanding data collections in every subject. In the following example, we will download data from the UC Irvine Machine Learning Repository (http://archive.ics.uci.edu/ml/index.php)and read them.

7.1.3 **Download Web data** Download the following .txt and .csv files format to your data folder and read them into Scilab environment and inspect check their sizes.

Solution:

- *fertilityDiagnosis.txt //*
  from http://archive.ics.uci.edu/
  ml/machine-learning-databases/00244/
- *winequality − red.csv //*
  from http://archive.ics.uci.edu/ml
  /machine-learning-databases/wine-quality/

```
//read the first 9 numerical variables
fertility =read('fertility_Diagnosis.txt',-1,9);
fertility(1:3,:)// first three rows
size(fertility)
//Notepad shows that the first row is the header and the separator is ;
header = 1;
Wine = csvRead("winequality-red.csv", ";", "double",[],[],[],[], header);
size(Wine)

--> fertility =read('fertility_Diagnosis.txt',-1,9);
--> fertility(1:3,:)// first three rows
  ans  =
  -0.33   0.69   0.   1.   1.   0.   0.8   0.    0.88
  -0.33   0.94   1.   0.   1.   0.   0.8   1.    0.31
  -0.33   0.5    1.   0.   0.   0.   1.   -1.    0.5
--> size(fertility)
```

```
  ans  =
     100.   9.
--> Wine = csvRead("winequality-red.csv", ";", "double",[],[],[],[], header);
--> size(Wine)
  ans  =
     1599.   12.
```
□

Data Analysis Scilab Module

Scilab has added a recent Module by Holger Nahrstaedt: rdatasets that can be added using the Scilab
ATOMS menu and includes over 700 data sets from the following r packages:

"datasets", "boot", "KMsurv", "robustbase", "car", "cluster", "COUNT", "Ecdat", "gap", "ggplot2",
"HistData", "lattice", "MASS", "nlreg", "plm", "plyr", "pscl", "reshape2", "rpart", "sandwich", "sem",
"survival", "vcd", "Zelig"

7.1.4 R Packages data sets
- Print out all loaded R packages
- Print out all data sets included in ggplot2 package
- read the data set geyser from the MASS package and print out the geyser variable definition and
  plot them

Solution:

```
// To list all loaded packages
rdata=rdataset_listgroups();
rdata'

// To list the data sets in ggplot2
rdataset_listsets("ggplot2")'
--> rdataset_listsets("ggplot2")'
 ans  =
!seals  presidential  msleep  mpg  movies  midwest  economics  diamonds  !

// To read geyser data set from MASS
[geyser,desc] = rdataset_read("MASS","geyser");
// read description
disp(desc);// to suppress the output
// read header of data
disp(geyser(1))
// Plot Duration over Waiting Time
clf
plot(geyser.waiting,geyser.duration,"bo")
xlabel("Waiting");ylabel("Duration");
```
□

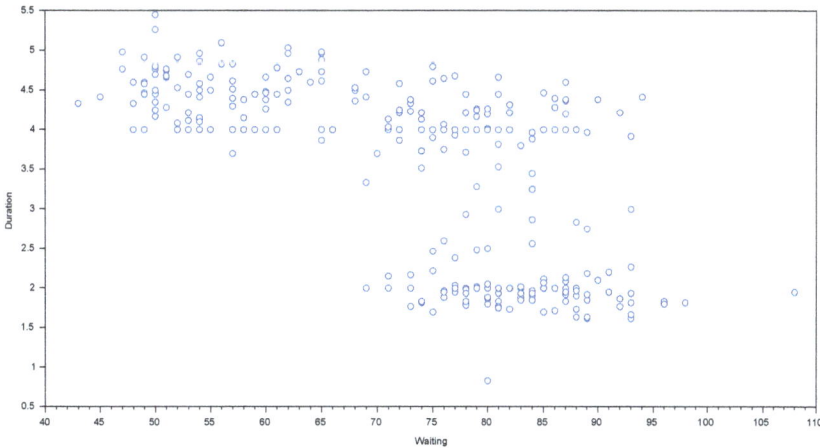

Figure 7.2: Geyser data set

7.2 Data Analysis and Statistical Functions

Scilab can accomplish a basic statistical analysis of numerical rectangular data types. The columns usually represent different variables and the rows represent different cases. Most functions have column and row options: The string "r" or 1 for columns and "c" or 2 for rows operations as will be demonstrated in the examples. We will select the most commonly used statistical functions in the following table. However, you need to check the Scilab help manual for a complete set of functions.

correl(x, y)	Correlation coefficient between two vectors x, y.
cov(A)	Covariance matrix of A .
gsort(A, option)	Sort the elements of A. options: 'd' decreasing, 'i' increasing order, in addition to 'r', 'c'.
mean(A, option)	Mean value of A, options:'r' or 'c'.
max(A, option)	Maximum component of A.
median(A, option)	Median value of A, options:'r' or 'c'.
min(A, option)	Minimum component of A.
perctl(A, y)	Percentile of A, y is an optional vector.
(a,b)=reglin(x,y)	Linear regression between x and y with regression coefficients.
sample(n,A)	Sample extracted from A with size n.
stdev(A, option)	Standard deviation of A.
strange(A, option)	Range of A.
sum(A, option)	Sum of elements of A A.
tabu(A, option)	Frequency of A with options 'i' increasing and 'd' decreasing.
variance(A, option)	Variance of A.

7.2.1 Find the correlation between the temperature of the three cities: NYC, Moscow, and Cairo introduced in the previous section and compute their average temperature.

Solution:

```
// We reload the data from the previous section
NewYork = [27 27 21 21 27 28 27 27 26 21 25 23 18 23 27]';
Moscow = [23 26 22 24 27 29 31 31 31 27 28 24 20 21 21]';
Cairo = [37 37 37 34 33 35 33 33 34 36 39 33 33 32 35]';
Temps = [NewYork, Moscow, Cairo]
//Use the commands
NYC_Moscow=correl(NewYork,Moscow)
NYC_Cairo=correl(NewYork,Cairo)
Moscow_Cairo=correl(Moscow,Cairo)

ave_three_cities=mean(Temps,'r')
//The output
--> NYC_Moscow=correl(NewYork,Moscow)
 NYC_Moscow  =
    0.5583509
--> NYC_Cairo=correl(NewYork,Cairo)
 NYC_Cairo  =
    0.1021188
--> Moscow_Cairo=correl(Moscow,Cairo)
 Moscow_Cairo  =
   -0.0030155
--> ave_three_cities=mean(Temps,'r')
 ave_three_cities  =
    24.533333    25.666667    34.733333
```

□

7.2.2 The following data are the scores of five students on three quizzes:

```
Q=[78,83,97;
92,90,100;
84, 53, 45;
52, 67, 71;
23, 48, 65];
```

Compute the max, min, range, and average score of the entire group, the average score on each quiz, and the average score of each student.

Solution:

```
//Quizes study
Q=[78,83,97;
92,90,100;
84, 53, 45;
52, 67, 71;
23, 48, 65];
```

```
Q
format(6)// fix the number of printed digits:default is 10
max(Q) // largest score
min(Q) // smallest score
strange(Q) // the range
mean(Q)// average score of all scores
mean(Q,'r') // average score on each quiz
mean(Q,'c') // average score of each student
//the output
--> Q
 Q  =
    78.    83.    97.
    92.    90.    100.
    84.    53.    45.
    52.    67.    71.
    23.    48.    65.
--> format(6)// fix the number of printed digits:default is 10
--> max(Q) // largest score
  ans  =
    100.
--> min(Q) // smallest score
  ans  =
    23.
--> strange(Q) // the range
  ans  =
    77.
--> mean(Q)// average score of all scores
  ans  =
    69.87
--> mean(Q,'r') // average score on each quiz
  ans  =
    65.8    68.2    75.6
--> mean(Q,'c') // average score of each student
  ans  =
    86.
    94.
    60.67
    63.33
    45.33
```

□

Sorting

The *gsort* function is used to sort and order data sets.

7.2.3 Define the matrix
$$A = \begin{pmatrix} 0 & 9 & 9 \\ 3 & 0 & 3 \\ 9 & 3 & 0 \end{pmatrix}$$

and apply *gsort* function with different options and observe the result

```
A=[0, 9, 9;
   3, 0, 3;
   9, 3, 0 ];
// ordering in descending order (default)
gsort(A)
--> gsort(A)
 ans  =
    9.   3.   0.
    9.   3.   0.
    9.   3.   0.
--> gsort(A,'c') // sorting in each row
 ans  =
    9.   9.   0.
    3.   3.   0.
    9.   3.   0.
--> gsort(A,'r') // sorting in each column
 ans  =
    9.   9.   9.
    3.   3.   3.
    0.   0.   0.
--> gsort(A,'g','i') //sorting in increasing order
 ans  =
    0.   3.   9.
    0.   3.   9.
    0.   3.   9.
--> gsort(A,'lr') // sorting based on the left column
 ans  =
    9.   3.   0.
    3.   0.   3.
    0.   9.   9.
--> gsort(A,'lr','i') // sorting in increasing order based on the left column
 ans  =
    0.   9.   9.
    3.   0.   3.
    9.   3.   0.
--> find(A>8) // identify the indices of A with values larger than 8,
// A is treated as one dimensional array
 ans  =
    3.   4.   7.
```

The Regression Equation

Predicting a variable based on another variable whenever the scatter plot appears to fit a straight line. This line is called the Least square line.

7.2.4 the following data are selected from 19 students where x is the first exam score out of 100 and y is the second exam score out of 100. Can you predict the second exam score of a random student if you know the first exam score?

Solution: *The exam data are listed for this group of students:*

```
exam1=[77, 82, 98, 74, 81, 71, 86, 71, 75, 100, 100, 88,...
       79, 98, 71, 98, 69, 77, 72];
exam2=[79, 22, 83, 85, 93, 93, 96, 26, 52, 100, 100, 90,...
       73, 87, 58, 100, 74, 45, 74];
correl(exam1,exam2)
--> correl(exam1,exam2)
 ans  =
    0.5220981
clf
x=exam1;y=exam2;
plot(x,y,'o')
[b1,b0]=reglin(x,y);
b0
b1
//y=-17.13 +1.12*x
plot(x,b0+b1*x,'r','thickness',3)
a=gca;//get the handle of the newly created axes
a.data_bounds=[60,0;110,110]; // viewing window
xlabel('Exam1')
ylabel('Exam2')
title('Exam Prediction: Scatter and Regression Line Plots ')
```

□

Percentiles for a data set: exam1

7.2.5 The ages for Academy Award winning best actors are given below. Arrange the list in order from smallest to largest. Use the percentile to show the quantile and use the quart function to show the quartiles and compare to the median.

```
age =[18; 18; 21; 22; 25; 26; 27; 29; 30;

 31; 31; 33; 36; 37; 37; 41; 42; 47; 52;

  55; 57; 58; 62; 64; 67; 69; 71; 72; 73;

  74; 76; 77]'
age_i=gsort(age,'g','i')// increasing order
median(age_i)// median
y=10:10:100
perctl(age_i,y)// percentiles
```

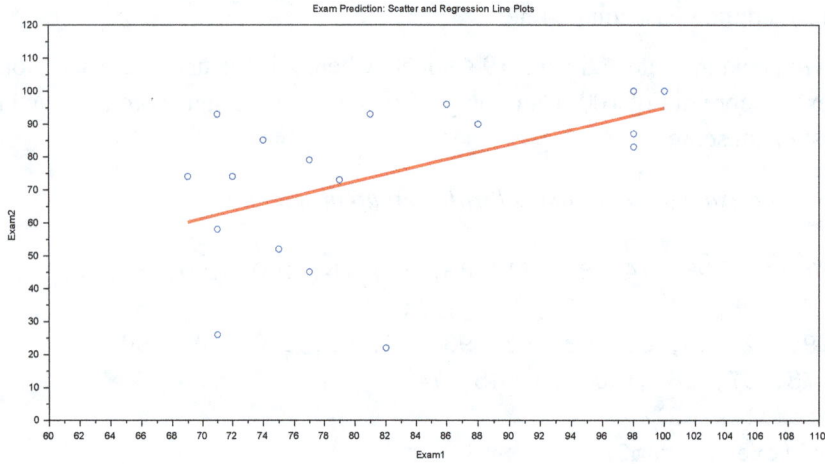

Figure 7.3: Scatter and Regression Exam Plots

```
quart(age_i)//quartiles

--> age_i=gsort(age,'g','i')// increasing order
 age_i  =

   18.   18.   21.   22.   25.   26.   27.   29.   30.   31.
   31.   33.   36.   37.   37.   41.   42.   47.   52.   55.
   57.   58.   62.   64.   67.   69.   71.   72.   73.   74.
   76.   77.
--> median(age_i)// median
 ans  =
   41.5
--> y=10:10:100
 y  =
   10.   20.   30.   40.   50.   60.   70.   80.
   90.   100.
--> perctl(age_i,y)// percentiles
 ans  =
   21.3   3.
   26.6   6.
   30.9   9.
   36.2   13.
   41.5   16.
   54.4   19.
   62.2   23.
   69.8   26.
   73.7   29.
```

```
   77.    32.
--> quart(age_i)//quartiles
 ans  =
   29.5
   41.5
   65.5
```

7.3 Data Visualization and Simulations

Scilab has many graph types for visualizing distributions, patterns, classifications, and trends: Scatter plots, histograms, bar, and pie plots. hopefully, it will be enhanced with additional types of data visualizations.

plot(x,y,options)	general line plot for numerical vectors x, y.
scatter(x, y, options)	Scatter lot with size of characters and colors for numerical vectors x, y.
scatter3(x, y, z, options)	3-d Scatter lot with size of characters and colors for numerical vectors x, y, z.
histplot(n, data, options)	histogram for data with n bins.
bar(x, y, options)	bar plots with options: width, color, horizontal and vertical bars.
pie(x, options)	pie plot for x with string options and separations.

7.3.1 Histograms

7.3.1 Generate a random vector of size length 1000, normally distributed with mean value of 65 and standard deviation of 9. Create a histogram of these data.

Solution:

```
n=1000;// length
Av = 65; // mean
Sd = 9; //standard deviation
x=grand(1,n, 'nor', Av, Sd);
clf
subplot(3,1,1)
histplot(15, x, style=5)
xtitle("Normalized histogram")
subplot(3,1,2)
histplot(20, x,normalization=%f, style=2)
xtitle("Count histogram")
subplot(3,1,3)
histplot(20, x, style=9,polygon=%t )
xtitle("Histogram with chart")
```

□

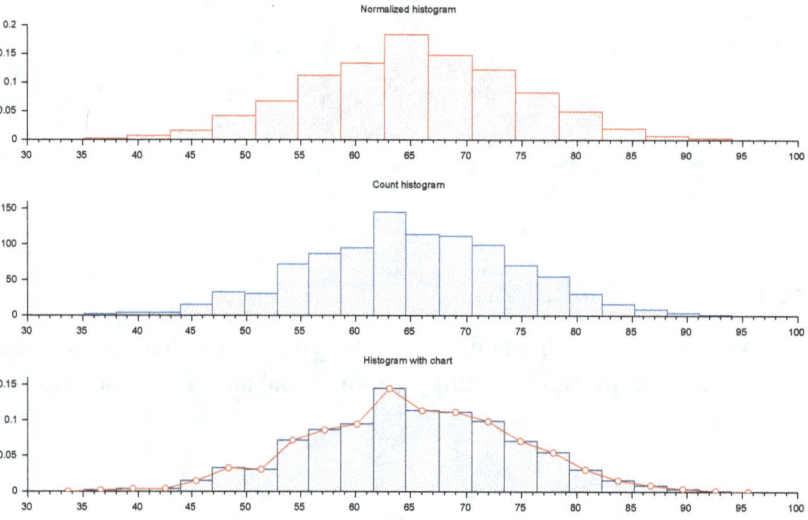

Figure 7.4: Histogram with Options

7.3.2 Scatter Plots

7.3.2 Load the data set airquality from the R datasets package. Describe the data sets and their variables.

> **Solution:** *We follow the commands and copy their output*

```
// Load the data set airquality
[airquality,desc] = rdataset_read("datasets","airquality");
// print out the names of the variables
disp(airquality(1))
--> disp(airquality(1))
!struct  Nr  Ozone  Solar_R  Wind  Temp  Month  Day  !
// print out the data info
desc
!Daily air quality measurements in New York, May to September 1973.
!A data frame with 154 observations on 6 variables.
!-  ''Ozone'': Mean ozone in parts per billion
 from 1300 to 1500 hours at Roosevelt Island
```

□

7.3.3 Create a scatter plot of Ozone versus temperature in the airquality data. Use the size and color option to improve the visualization of the scatter plot.

> **Solution:**

```
clf
subplot(2,1,1)
scatter( airquality.Temp,airquality.Ozone,"fill")
```

```
xtitle('Scatter plot of Temperature and

 Ozone', 'Temperature', 'Ozone');
// set color map
gcf().color_map = hotcolormap(64);
// colors according to x values
s = airquality.Temp;
c=airquality.Ozone;
subplot(2,1,2)
scatter( airquality.Temp,airquality.Ozone,s,c,
"fill");
xtitle('Scatter plot of Temperature (size)

 and Ozone (color)', 'Temperature', 'Ozone');
```

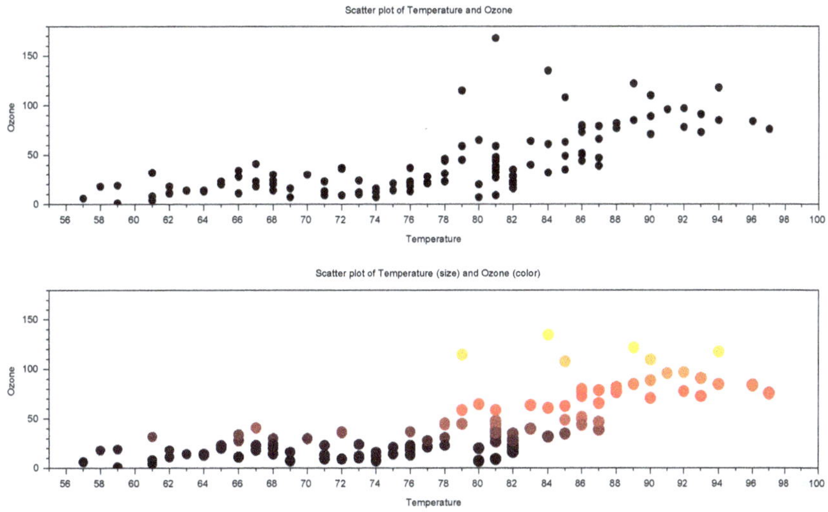

Figure 7.5: Scatter plot

7.3.3 Bar Plots

7.3.4 Create a bar graph for the distribution of grades in a classroom:

A	B	C	D	F
2	4	6	3	1

Solution:

```
clf
grade_freq=[ 2 4 6 3 1];
subplot(1,3,1)
bar(grade_freq)
subplot(1,3,2)
bar(grade_freq,0.5,'green');
a = gca();
a.x_ticks.labels = ['A'; 'B'; 'C';'D';'F'];
xtitle('Grade Distribution', 'Grades', 'Frequency');
subplot(1,3,3)
barh(grade_freq,0.5,'magenta');
a = gca();
a.y_ticks.labels = ['A'; 'B'; 'C';'D';'F'];
xtitle('Grade Distribution',  'Frequency','Grades',);
```

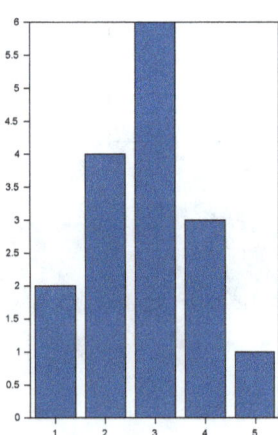 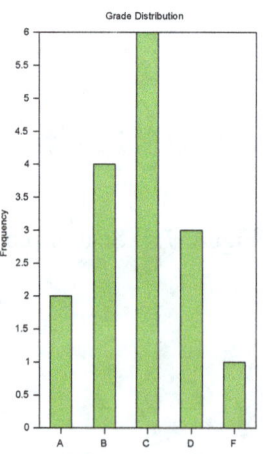 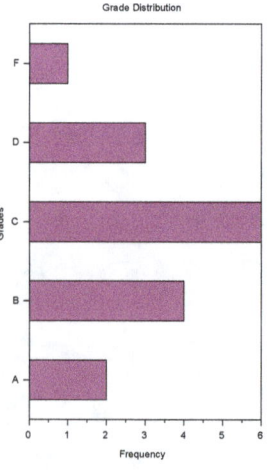

Figure 7.6: Bar graphs

7.3.4 Pie Charts

7.3.5 Create a pie chart for the distribution of grades in a classroom from last example.:

Solution:

```
clf
subplot(2,1,1)
pie(grade_freq)
title('Pie Chart','color','red','fontsize',2);
subplot(2,1,2)
pie(grade_freq,[1,0,1,0,0],['A'; 'B'; 'C';'D';'F']);
title('Course Grade Distribution','color','blue','fontsize',3);
```

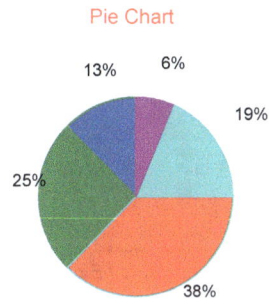

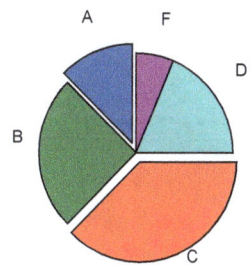

Figure 7.7: Pie charts

7.3.6 Create a pie chart for US Federal Budget expenditures for 2019 in billions of dollars:

Pensions	1111.2
Protection	41.6
Health Care	1252.2
Education	156.1
General Government	58.6
Defence	939.4
Welfare	378.6
Transportation	98.9
Interest	393.5
Other Spending	92.5

Solution:

```
//US Federal Budget for 2019 in billions of $
Expenditures=[1111.2; 41.6; 1252.2;156.1;
 58.6; 939.4; 378.6;  98.9;393.5;  92.5];

names=['Pensions'; 'Protection'; 'Health
 Care'; 'Education'; 'General
   Government';...
'Defence'; 'Welfare'; ...

 'Transportation'; 'Interest'; 'Other
   Spending'];
clf
subplot(1,2,1)

pie(Expenditures,[1,0,1,0,0,1,0,0,0,0])

title('US Federal Budget for 2019 in billions of
 dollars','color','red','fontsize',4);

subplot(1,2,2)
pie(Expenditures,[1,0,1,0,0,1,0,0,0,0],names)
title('US Federal Budget for 2019 in
 billions of
   dollars','color','green','fontsize',4);
```

□

US Federal Budget for 2019 in billions of dollars

US Federal Budget for 2019 in billions of dollars

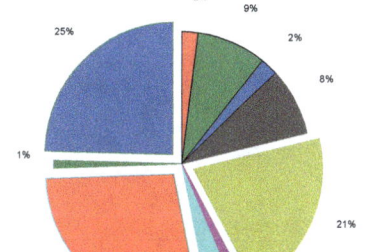

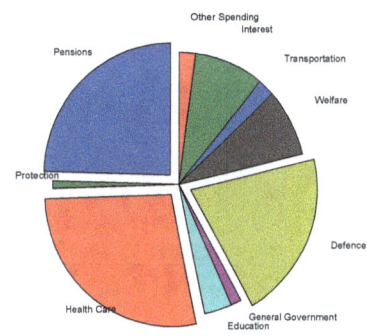

Figure 7.8: Pie chart for US Budget

Chapter 8

Computational Applications

Scilab is a powerful tool in exploring and researching numerical investigation and mathematical explorations such as optimization problems, numerical procedures in differential and integral calculus, Fourier analysis and imaging, differential equations and dynamical systems, in linear algebra to name a few applications. In the following sections, we will show how to implement Scilab functions to explore some of these problems.

8.1 Calculus

Scilab has many built-in functions to perform numerical differentiation and integration with many options.

8.1.1 Differentiation

The linear approximation for first-order derivative of a given function $f(x)$ is given by

$$f'(x) \approx \frac{f(x+h) - f(x)}{h}$$

which is based on the following definition for small h:

$$f'(x) = \frac{df}{dx} = \lim_{h \to o} \left(\frac{f(x+h) - f(x)}{h} \right)$$

The command `diff` performs the first finite difference between the elements of any vector y:

$$\Delta y = y_{i+1} - y_i$$

8.1.1 Plot the function $y = sin(x)$ and its first derivative using the numerical differences *diff* on the interval $[0, 2\pi]$.

Solution:

```
clear
x=0:0.01:2*%pi;
y=sin(x);
```

```
plot(x,y)
dydx=diff(y)./diff(x);
plot(x,y,x(1:$-1),dydx,':','thickness',2);
legend(["sin(x))", "cos(x)"]);
title('Function with its numerical derivative')
```

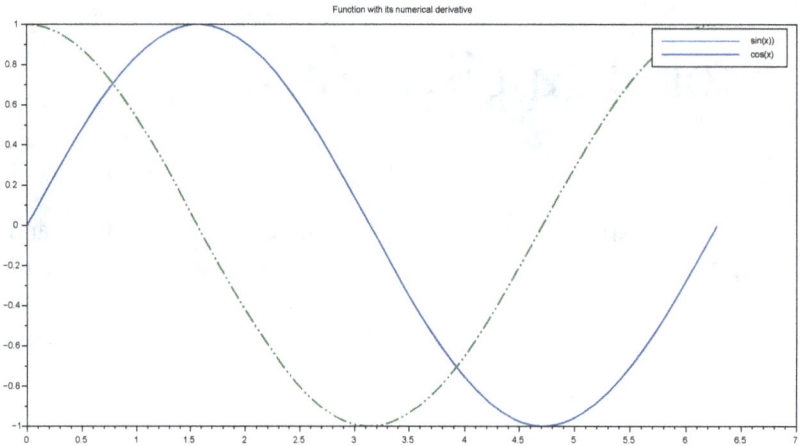

Figure 8.1: y=sin(x) and y=diff(y)/diff(x)
,labelfig7.1

8.1.2 Plot the function $y = x^3 - x$ on the interval $[-1.5, 1.5]$ with 8 and 16 points and the corresponding first order derivatives using the `diff` function with bar graphs on 4 by 4 subplots.

Solution:

```
// date and name
clear
x8=linspace(-1.5,1.5,8);
x16=linspace(-1.5,1.5,16);
y8=x8.^3-x8;
y16=x16.^3-x16;
slope8=diff(y8)./diff(x8);
slope16=diff(y16)./diff(x16);
x7=x8(1:7); //Adjust the x values for diff
x15=x16(1:15);
subplot(2,2,1)
plot(x8,y8,'-o')
xtitle('y=x^3-x')
subplot(2,2,2)
```

```
bar(x7,slope8)
xtitle('Slope of y=x^3-x' )
subplot(2,2,3)
plot(x16,y16,'-o')
xtitle('y=x^3-x')
subplot(2,2,4)
bar(x15,slope16)
xtitle('Slope of y=x^3-x' )
```
□

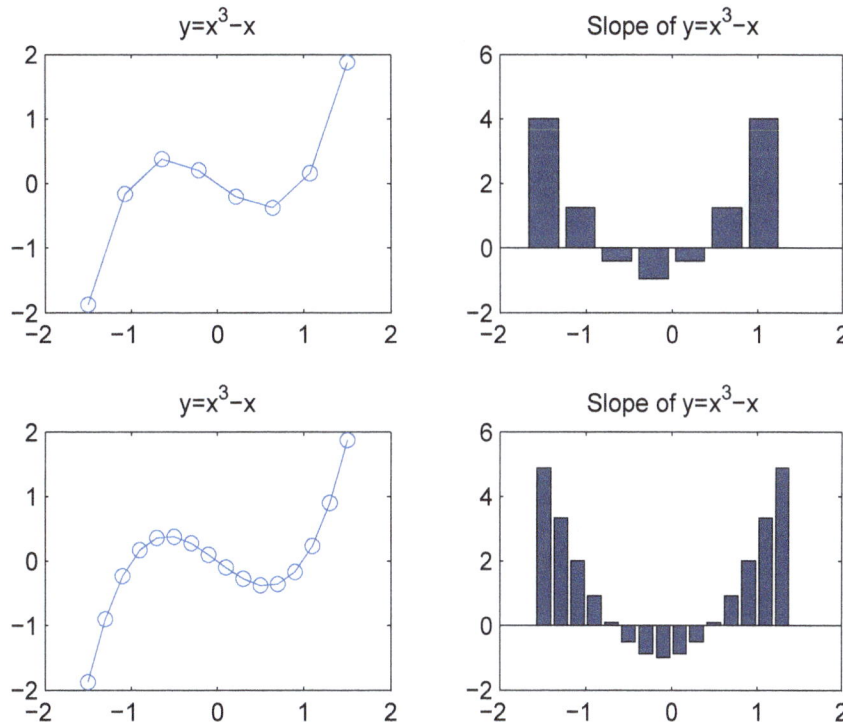

Figure 8.2: Example 1

8.1.3 Plot the function $y = x\sin(x)$ on the interval $[-\pi, \pi]$ with 8 and 16 points and the corresponding derivatives using the `diff` function with bar graphs on 4 by 4 subplots.

8.1.2 Numerical Integration

We can evaluate definite integrals of the form

$$I = \int_a^b f(x)dx$$

by methods of numerical integration which are called numerical quadratures, such as trapezoidal, Simpson formulas, and others. Scilab has a set of built-in functions to calculate the definite integrals;

intg(a, b, f)	evaluates the definite integral of a single variable function f over the interval $[a, b]$
intd2d(X,Y,f)	evaluates the double integral of a function f over 2D domain and X, Y are matrices of the vertices of the triangulation of the domain.
intd3d(X,Y,Z,f)	evaluates the 3D integral of a function f over 3D region and X, Y, Z are matrices of the vertices of the tetrahedron of the region.

The construction of the higher dimensional domains is not simple.

8.1.4 Find

$$I = \int_0^2 (3x^2 - 2x + 5)dx$$

Solution: *We know that the exact value of this integral is $I = x^3 - x^2 + 5x|_0^2 = (8 - 4 + 10) - (0) = 14$. The Scilab evaluation of this integral is entered as follows:*

```
function y=f(x),y=3*x^2-2*x+5;endfunction
exact=14;
I=intg(0,2,f)
abs(exact-I)
```

$\square$

8.1.5 Find

$$I_1 = \int_0^1 \cos(\pi t^2)dt$$

Solution:

```
function y=f(x),y=cos(%pi*x^2);endfunction
I=intg(0,1,f)

-->I=intg(0,1,f)
 I  =
     0.3739828
```

$\square$

8.1.6 Compute the following two integrals and compare the result of the two functions by requesting the long format
1. $q1 = \int_0^2 (\sqrt[3]{x + x^2})dx$
2. $q2 = \int_0^4 x \ln(x)dx$

Solution: *We can find the integral directly on the command window as follows*

```
function y=f(x),y=(x+x^2)^(1/3);endfunction
I=intg(0,2,f)
-->I1=intg(0,2,f)
 I1  =
      2.4175887
-->[I, e]=intg(0,2,f)
```

```
e   =
    1.885D-09
 I   =
    2.4175887
```

```
function y=f(x),y=x*log(x);endfunction
I2=intg(0,4,f)
-->I2=intg(0,4,f)
 I2   =
    7.0903549
```

□

8.1.7 Plot the function $|\frac{sin(10x)}{x}|$ over the interval $[0,2]$ and find the area under this curve. Try to experiment with the convergence or divergence of this integral as you increase the upper bound toward infinity.

Solution:

```
function y=f(x),y=abs(sin(10*x)/x) ;endfunction
I=intg(0,2,f)
-->I=intg(0,2,f)
 I   =
    3.0128756
```

```
x=linspace(%eps,2,200);
fplot2d(x,f)
xtitle('|sin(10*x)./x|')
```

□

8.1.8 Approximate the area under the graph of $y = 4 - x^2$ and bounded by $x = 0, x = 2$ and the y-axis.

8.1.9 Compute the double integral $\int_0^1 \int_0^1 (yx^2 + xy^2)dydx$

Solution: *We need to triangulate the square into two triangles: ABC with vertices $A(0,0), B(1,0), C(1,1)$ and ACD with vertices $A(0,0), C(1,1), D(0,1)$. The double integration requires the construction of two matrices $X(3,2)$ of the abscissa of the vertices and $Y(3,2)$ of the ordinates of the vertices:*

```
function z=f(x,y);z=y*x^2+x*y^2 ;endfunction
```

```
X=[0,0;1,1;1,0]; // Abscissa of the vertices of ABC and ACD
Y=[0,0;0,1;1,1]; // Ordinates  of the vertices of ABC and ACD
//deff('z=f(x,y)','z=cos(x+y)')
[I,e]=int2d(X,Y,f)
// computes the integrand over the square [0 1]x[0 1]
-->[I,e]=int2d(X,Y,f)
```

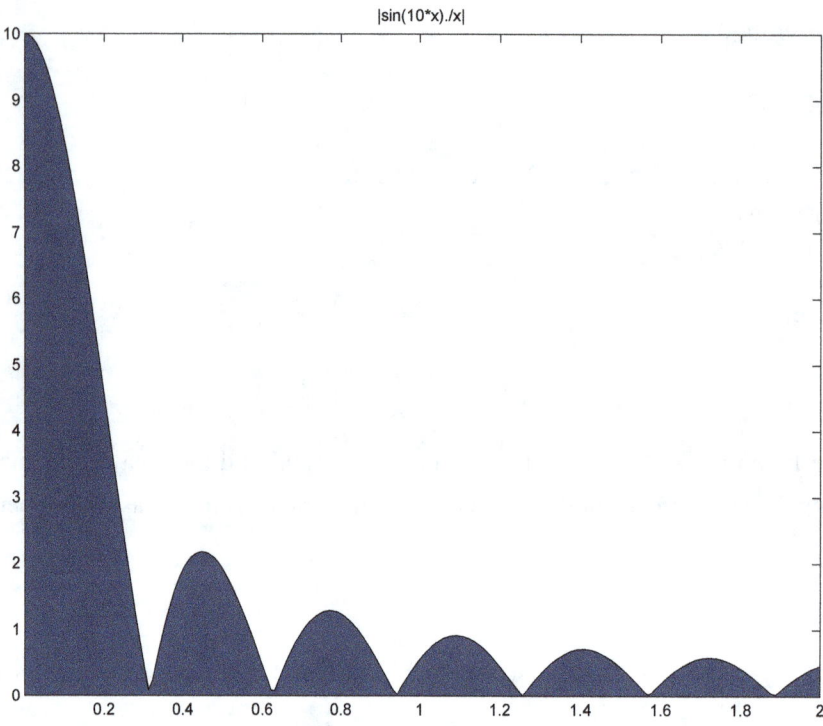

Figure 8.3: Area under a curve

```
e   =
    7.401D-17
I   =
    0.3333333
```

□

8.1.10 Compute the double integral $\int_\pi^{2\pi}\int_0^\pi (y\sin(x)+x\cos(y))\,dy\,dx$

Solution: *We need to triangulate the rectangle into two triangles: ABC with vertices*
$A(\pi,0), B(2\pi,0), C(2\pi,\pi)$ *and ACD with vertices* $A(\pi,0), C(2\pi,\pi), D(\pi,\pi)$. *The double*
integration requires the construction of two matrices $X(3,2)$ *of the abscissa of the vertices*
and $Y(3,2)$ *of the ordinates of the vertices:*

```
x=linspace(%pi,2*%pi,30);
y=linspace(0,%pi,30);
function z=f(x,y)
    z=y*sin(2*x)+x*cos(3*y)
endfunction

clf
```

```
fplot3d(x,y,f)

//domain vertices
X=[%pi,%pi;2*%pi,2*%pi;2*%pi,%pi];
Y=[0,0;0,%pi;%pi,%pi];

[I,e]=int2d(X,Y,f)
// the output
-->[I,e]=int2d(X,Y,f)
 e  =
    9.810D-11
 I  =
   - 9.8696044
```

□

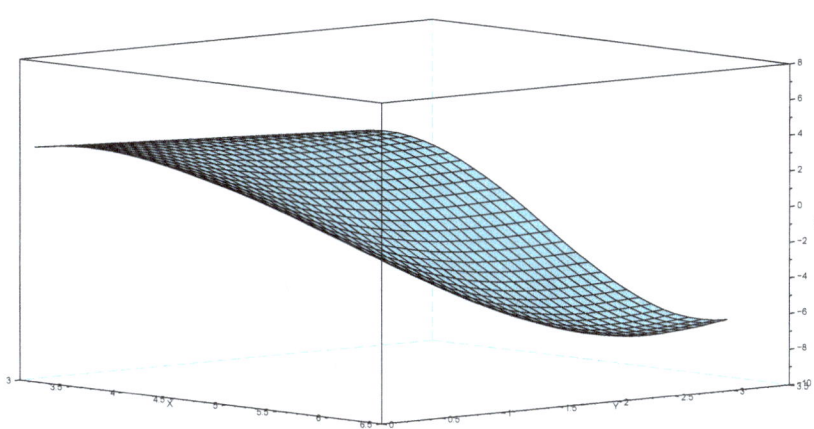

Figure 8.4: Double Integral

8.1.3 Zeros of equations

We can solve algebraic equation $f(x) = 0$ by applying the function
```
 fsolve(x_0,'equation') //x_0 initial guess
```
8.1.11 Find the solution of $sin(x) = e^x - 5$

> **Solution:**   *We graph the equation $y = sin(x) - e^x + 5$, from the plot we guess the solution to be $x = 1$ corresponding to zero graph height. Use this value in the equation solver.*

```
x=-2:.01:2;y=sin(x)-exp(x)+5;plot(x,y)
function y=f(x);y=sin(x)-exp(x)+5 ;endfunction
plot(x,f)
```

```
plot(x,0*x, "r")// x-axis
title('Zeros are intersection of f(x) with x-axis','fontsize',3)

x1=fsolve(1,f)
-->x1=fsolve(1,f)
 x1  =
    1.7878415
```

$\square$

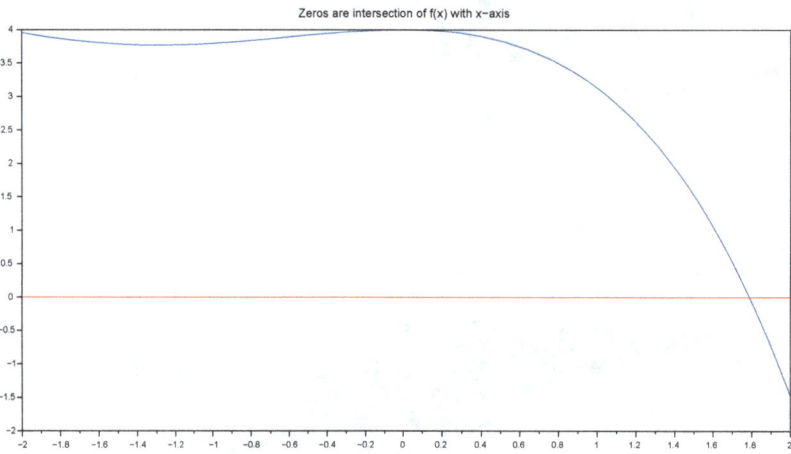

Figure 8.5: Zeros of function

8.1.12 Find the solution of the nonlinear algebraic equation $x^2 = 1 - \tan(x)$.

Solution: *We plot the equation $y = x^2 + \tan(x) - 1$ over $[-1, 1]$, from the graph we guess the initial solution to be 0.5 which is used in the solution solver.*

```
x=-1:.01:1;y=x.^2+tan(x)-1;plot(x,y)
function y=f(x);y=x^2+tan(x)-1; ;endfunction
x1=fsolve(0.5,f)
-->x1=fsolve(0.5,f)
 x1  =
    0.5832485
```

$\square$

8.1.13 Find all the real roots of the function $f(x) = x^3 - 4x^2 - 0.5x + 3.8$.

8.1.14 Find all the real roots of the function $g(x) = e^x - x - 1.9$.

8.2 Differential Equations

The Scilab has many differential equations solvers, including *bvode* boundary value problems for ODE using collocation method, *dae* Differential algebraic equations solver, *ode* ordinary differential equation solver, and few more. In this section, we focus on the common solver *ode* and its applications.

The Scilab solver *ode* calculates the solution of first-order differential equation with given initial conditions. that is a solver of the initial value problems of the form

$$\frac{d}{dt}y(t) = f(t, y(t)), \quad y(t_0) = y_0$$

which is a vector notation for the set of differential equations

$$\dot{y}_1 = f_1(t, y_1, y_1, \cdots, y_n), \quad y_1(t_0) = y_{10}$$
$$\dot{y}_2 = f_2(t, y_1, y_1, \cdots, y_n), \quad y_2(t_0) = y_{20}$$
$$\vdots$$
$$\dot{y}_n = f_n(t, y_1, y_1, \cdots, y_n), \quad y_n(t_0) = y_{n0}$$

where $\dot{y}_i = dy_i/dt$, n is the number of first-order equation, and y_{i0} is the initial condition associated with the ith equation. when an initial value problem is written in higher order differential equation, it must be rewritten as first order. For example, the pendulum equation:

$$x'' + kx' + sin(x) = 0$$

To transform this equation to a system of first-order equations, we choose $y_1 = x$ and $y_2 = x'$, then the pendulum equation becomes

$$\dot{y}_1 = y_2$$
$$\dot{y}_2 = -ky_2 - sin(y_1)$$

The syntax of the Scilab *ode* solver has the form

$$y = ode(y_0, t_0, t, f(t, y))$$

The function *ode* has many options which are fully listed in Scilab Help.

The main steps in Applying the ODE solver are
1. Solve for the derivative to identify the right-hand side of the ode, that is $f(t, y)$
2. Define a function for the right-hand side f
3. Define the interval for the solution: t
4. Specify the initial conditions t_0, y_0
5. Call the ODE solver and save the output: $y = ode(y_0, t_0, t, f)$
6. Plot the solution
7. Plot a vector field or phase plane as requested

First Order Differential Equations

8.2.1 Solve the differential equation

$$y' = -y\sin(2y) - \cos(x); \qquad y(0) = 1; \qquad x \in [0, 20]$$

Plot your solution.

Solution:

1. *Define the function of the right side:* $f(x, y) = -y\sin(2y) - \cos(x)$
2. *Vectorize the solution range:* x
3. *Identify the initial data:* $x0 = 0, y0 = 1$
4. *Call the ODE solver* ode(y0, x0, x, f) *and plot as follows*

```
function ydot=f(x,y); ydot=-y*sin(2*y)-cos(x); endfunction
x=0:0.05:20;
x0=0.0;
y0=1.0;
y=ode(y0,x0,x,f); plot(x,y,'b', 'thickness',3)
title (' Solution of dy/dx=ysin(2y)-cos(x)','fontsize', 3)
```

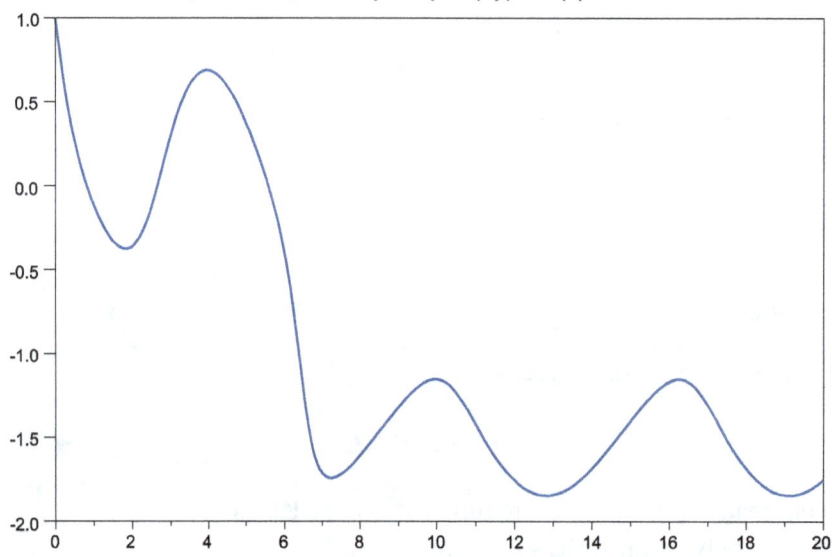

Figure 8.6: Example 1

□

8.2.2 Plot the solutions of the differential equation $\dfrac{dy}{dt} = y(1-y) - sin(6t)$

on $t \in [0, 5]$ for the following initial conditions

1. $y(0) = 0.2$
2. $y(0) = 0.5$
3. $y(0) = 1.0$
4. $y(0) = 1.5$

and draw a vector field using the *champ* command.

Solution: *Copy the following script to a file named* `ode1_solver.sce` *and run it.*

```
// Script file to solve a single first order DE
// with initial values
// dy/dt = y (1-y) -sin(6t) on 0<t<5
function ydot=f(t,y);
     ydot=y*(1-y)-sin(6*t);
endfunction
t=0:0.05:5;
t0=0;
y1=0.2;
y2=0.5;
y3=1.0;
y4=1.5;

y=ode(y1,t0,t,f); plot(t,y,'r')
y=ode(y2,t0,t,f); plot(t,y,'g')
y=ode(y3,t0,t,f); plot(t,y,'b')
y=ode(y4,t0,t,f); plot(t,y,'m')

/////////////////////
//clf
t=0: .2: 5; y=-.5:.2:2;
function z=ft(t,y),z=1;endfunction
function z=fy(t,y),z=y*(1-y)-sin(6*t);endfunction
zt=feval(t,y,ft); zy=feval(t,y,fy);
champ1(t,y,zt,zy)

xtitle('dy/dt= y(1-y)-sin(6t)')
```

□

8.2.3 Plot the solution of the first order, nonlinear differential equation

$$y' = \frac{y^3 sin(3x)}{x^2}$$

with initial condition $y(1) = 2$ on the interval $[1, 10]$.

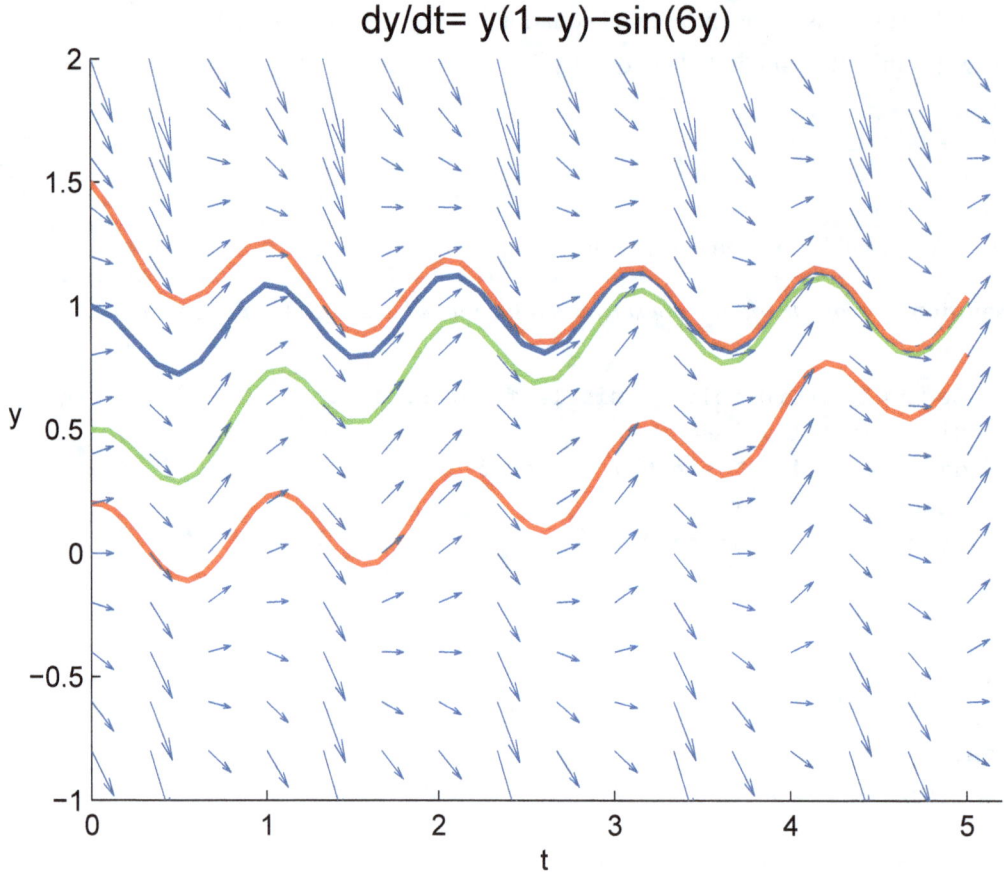

Figure 8.7: Example 2

Second Order Differential Equations

Higher-order differential equations must be reduced to a system of first-order differential equations. This can be achieved by introducing a set of new variables.

Reduction of higher-order equations to first-order systems

To write the second order equation

$$x'' + x = \sin t$$

We introduce new variables $y_1 = x, y_2 = x'$ and solve for $x'' = -x + \sin t$, this will be rewritten as first-order system as follows

$$y_1' = y_2$$
$$y_2' = -y_1 + \sin t$$

8.2.4 Express the second order equation as a first-order system of equation

$$2y'' + 6y' - 8y = 9\cos 2t$$

Solution: *Introduce new variables $y_1 = y, y_2 = y'$ and solve for $y'' = -3y' + 4u_1 +$*
$4.5\cos 2t$, so that

$$y_1' = y_2$$
$$y_2' = -3y_2 + 4y_1 + 4.5\cos 2t \qquad \square$$

8.2.5 Express the second order equation as a first-order system of equation

$$x''' - x^2 x'' + 5x'^3 = \sin t$$

Solution: *Introduce new variables $y1 = x, y_2 = x', y_3 = x''$ and solve for $x''' = x^2 x'' -$*
$5x'^3 + \sin t$, so that

$$y_1' = y_2$$
$$y_2' = y_3$$
$$y_3' = y_1^2 y_3 - 5y_2^3 + \sin t \qquad \square$$

8.2.6 Plot the solution of the damped motion

$$x'' + x' + 4x = 0$$

with initial condition $x(0) = 2, x'(0) = 3$ over the interval $[0, 12]$ and draw the phase plane with vector field.

Solution:

We introduce new variables $y_1 = x, y_2 = x'$ and solve for $x'' = -x' - 4x$, this will be rewritten as first-order system as follows

$$y_1' = y_2$$
$$y_2' = -y_2 - 4y_1$$

```
// Script file to solve a 2 by 2 first order system of DE
// with initial values
// x"=-x'-4x on 0<t<12

function [udr]=damp(t,y);
    ud1=y(2); ud2=-y(2)-4*y(1);
    udr=[ud1;ud2];
endfunction
t=0:.1:12;
t0=0;
u1=[2;3];
[sol]=ode(u1,t0,t,damp);
subplot(2,1,1)
plot(t,sol(1,:),'b','thickness',3)
xtitle('The position of motion','Time', 'x(t)')
```

```
subplot(2,1,2)
plot(sol(1,:),sol(2,:),'r','thickness',3)
xtitle('Phase Plane', 'Position', 'Velocity')

z1 = linspace(-1.5,2.5,15);
z2 = linspace(-4,3,15);
fchamp(damp,0,z1,z2) // Draw vector field
```

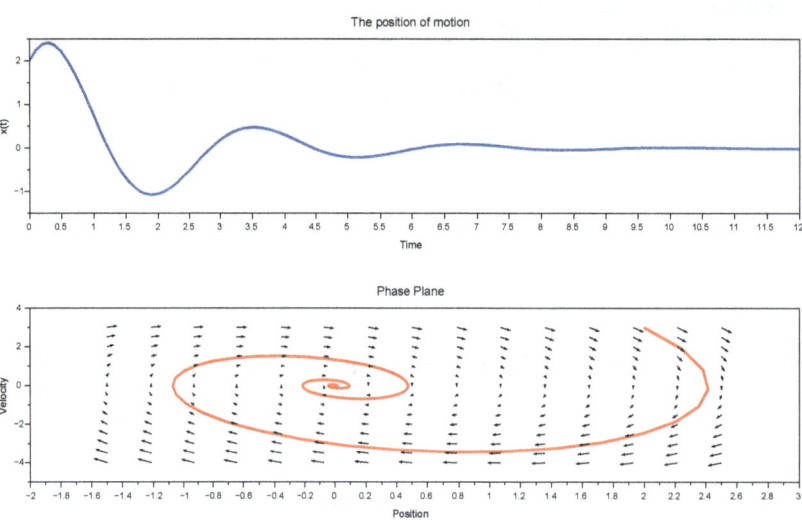

Figure 8.8: Damped motion

8.2.7 Plot the solution of the forced motion

$$x'' + x = \sin t$$

with initial condition $x(0) = 2, x'(0) = 1$ over the interval $[0, 10]$ and draw the phase plane with vector field.

8.2.8 Write a code to plot three solutions of the pendulum equation

$$\frac{d^2 y(t)}{dt^2} + 0.25 \frac{dy(t)}{dt} + \sin(y) = 0; \qquad t \in [0, 20]$$

and draw a vector field.

Solution: *We rewrite the second order equation as a system of first order equations by introducing new variables*

$$y1 = y, \qquad y2 = \frac{dy}{dt}, \qquad \frac{dy2}{dt} = \frac{d^2 y(t)}{dt^2}$$

as follows:

$$\dot{y}_1 = y_2; \qquad \dot{y}_2 = -0.25y_2 - \sin(y_1)$$

```
// Script file to solve a 2 by 2 first order system of DE
// with initial values
// y'' = -.25y'-sin(y) on 0<t<20


function [ydot]=pend(t,y);
    ydot1=y(2);ydot2=-0.25*y(2) -sin( y(1));
    ydot=[ydot1; ydot2 ]; endfunction

t=0:.2:20;
x0=[-5 ; -2]; t0=0;
[sol]=ode(x0,t0,t,pend);
plot(sol(1,:), sol(2,:), 'r' )
x2=[5;-2];
[sol]=ode(x2,t0,t,pend);
plot(sol(1,:), sol(2,:), 'g' )
x4=[-4.;3];
[sol]=ode(x4,t0,t,pend);
plot(sol(1,:), sol(2,:), 'b' )

x5=[4;-3];
[sol]=ode(x5,t0,t,pend);
plot(sol(1,:), sol(2,:), 'm' )

xf=linspace(-8, 8, 20); yf=linspace(-3, 3, 10);
fchamp(pend, 0, xf, yf);

xtitle('Pendulum: Phase Diagram with 4 trajectories', 'Position', 'Velocity')
```

□

The Lorenz Equations

The Lorenz equations for the functions $x(t), y(t)$ and $z(t)$ are given by

$$\dot{x} = \sigma(y - x)$$
$$\dot{y} = rx - y - xz$$
$$\dot{z} = xy - bz$$

These equations contain three parameters: σ, r, b. The most common values used in the Lorenz studies are $\sigma = 10, b = 8/3$ and $r = 28$. with a suitable step size $h = 0$ and using Euler solvers we plot the x plot and the three-dimensional plot.

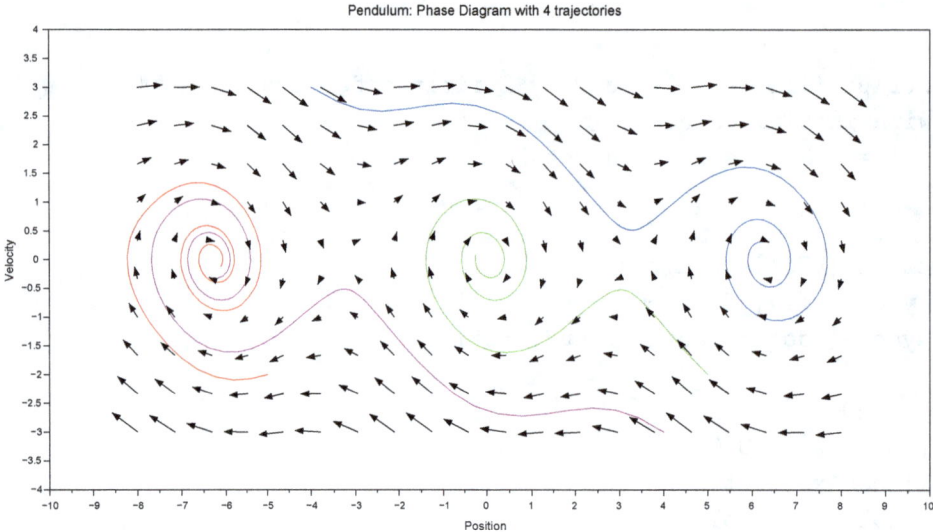

Figure 8.9: The Pendulum phase diagram

```
sig=10.0; b=8/3; r=20; // Parameters
t(1)=0.0; // Initial t
x(1)=0.1; y(1)=0.1; z(1)=0.1; // Initial x,y,z
dt=0.005; // Time step
nn=10000; // Number of time steps
for k=1:nn // Time loop
fx=sig*(y(k)-x(k)); // RHS of x equation
fy=-x(k)*z(k)+r*x(k)-y(k); // RHS of y equation
fz=x(k)*y(k)-b*z(k); // RHS of z equation
x(k+1)=x(k)+dt*fx; // Find new x
y(k+1)=y(k)+dt*fy; // Find new y
z(k+1)=z(k)+dt*fz; // Find new z
t(k+1)=t(k)+dt; // Find new t
end
clf
plot(t,x)
xtitle('Lorenz Solution of x(t)', 't', 'x')
clf
param3d(x,y,z')
e=gce();
e.foreground=color('gold');
e.thickness=3;
title('Lorenz Solution)
```

```
a=gca();
a.axes_visible="off";
a.box="off";
a.x_label.text =" ";
a.y_label.text =" ";
a.z_label.text =" ";
```

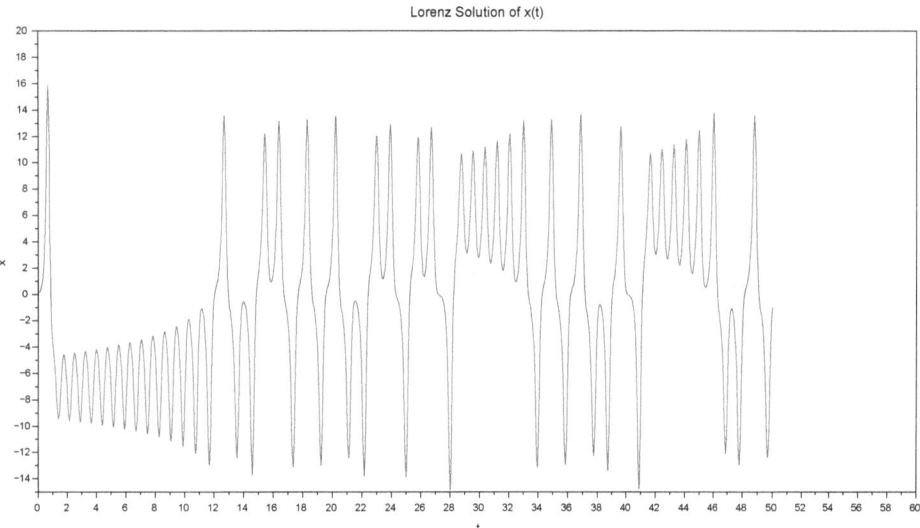

Figure 8.10: Lorenz Solution

8.3 Fourier Series

The Fourier series is an expansion of periodic functions in terms of trigonometric sines and cosines. If a function $f(x)$ is defined on an interval $[-p, p]$ and periodic with period $2L$, then the function can be expressed

$$f(x) = \frac{1}{2}a_0 + \sum_{m=1}^{\infty} a_m \cos(\frac{n\pi}{p}x) + \sum_{m=1}^{\infty} b_m \sin(\frac{n\pi}{p}x)$$

where n is any integer and a_m, b_m are the Fourier coefficients. The expressions for the coefficients are

Figure 8.11: Three-dimensional Lorenz Solution

$$a_n = \frac{1}{p} \int_{-p}^{p} f(x) \cos(\frac{n\pi}{p}x) dx$$

$$b_n = \frac{1}{p} \int_{-p}^{p} f(x) \sin(\frac{n\pi}{p}x) dx$$

$$a_0 = \frac{2}{p} \int_{-p}^{p} f(x) dx$$

8.3.1 Compute the Fourier expansions with 11 terms for the function $f(x)$ on the interval $[-\pi, \pi]$:

$$f(x) = |x| = \begin{cases} 0 & -pi \le x \le 0 \\ \pi - x & 0 < x \le \pi \end{cases}$$

Plot the function and the its Fourier approximation.

Solution:

```
// Define a function f(x) on an interval [-p,p]
p=%pi;

function y=f(x),
    if x<=0 then
```

```
            y=0;
    else
            y=%pi-x;
    end
endfunction
xv=-p:.001:p;plot(xv,f, 'thickness',2)

k=11;//Number of terms
a=zeros(1,k);b=a;
a0=intg(-p,p,f)/p;
for n=1:k
    np=n*%pi/p;
    a(n)=integrate('f(x)*cos(np*x)','x',-%pi,%pi)/p;
    b(n)=integrate('f(x)*sin(np*x)','x',-%pi,%pi)/p;
  end
xv=-3*p:.001:3*p; // to show larger interval
SUMF=a0/2;
for n=1:k
    np=n*%pi/p;
    SUMF=SUMF+a(n)*cos(np*xv)+b(n)*sin(np*xv);
end
plot(xv,SUMF,'r','thickness',2)
title("Fourier Expansion with k=11 terms")
```

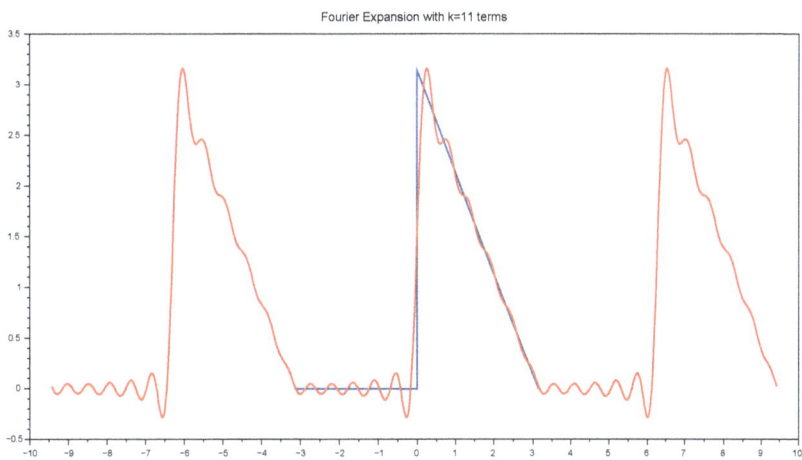

Figure 8.12: The Fourier Approximation with 11 terms

Scilab has the basic discrete fast fourier transform function *fft* and its inverse *ifft*. For example, the expansion of the function $y(t) = e^{13(i\pi/L)x}$ in terms of complex exponential should produce a single term. The following code graphs $|z|$ as a function of n, where z is the Fourier transform of y:

```
n=63; L=2;
t=-L:2*L/n:L;
y=exp(13*%i*%pi*t/L);
z=fft(y);
plot(0:63,abs(z),'magenta','thickness', 2)
title("Fourier transform of a single exponential")
```

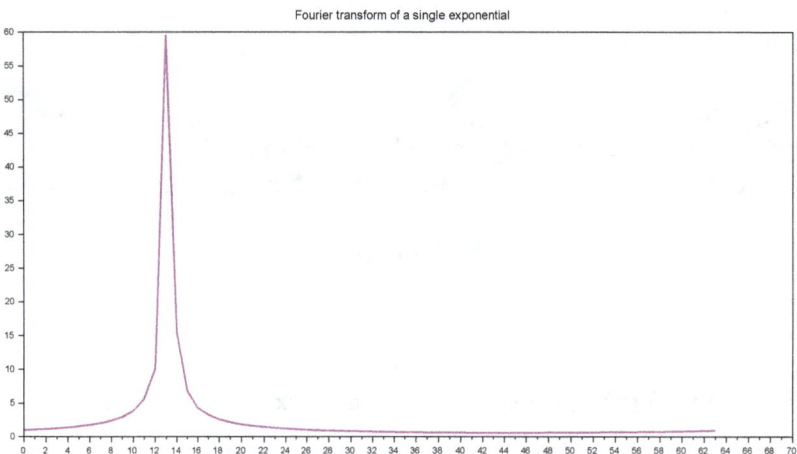

Figure 8.13: The absolute magnitude of the Fourier transform of a single exponential

If the function y were sine or cosine, then there would be two spikes as seen in the figure next.

```
n=63; L=2;
t=-L:2*L/n:L;
y=cos(13*%pi*t/L);
z=fft(y);
plot(0:63,abs(z),'red','thickness', 2)
title("Fourier transform of the cosine")
```

8.4 Polynomials

Scilab provides a collection of functions for manipulating polynomials. The polynomial $2x^3 - 5x^2 + 6x - 7$ can be represented by the vector $[2, -5, 6, -7]$. Derivatives, integrals, roots, and residues are some of these functions.

8.4.1 Use Scilab to build the polynomial $x^2 - 5x + 6$.

Solution:

```
// We can build the polynomial directly
```

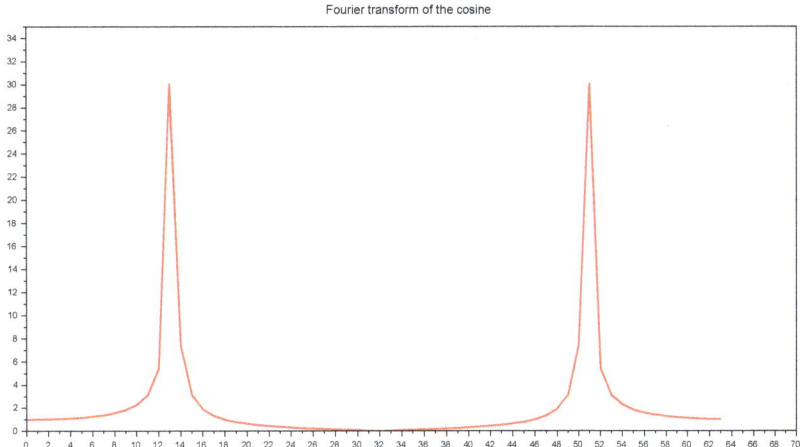

Figure 8.14: The absolute magnitude of the Fourier transform of a single cosine function

```
x = poly(0, "x");
p1=x^2 -5*x + 6
// We can use poly() with coefficients:
p2=poly([6 -5 1],"x", "coeff")
--> p2=poly([6 -5 1],"x", "coeff")
```

$$p2 = 6 - 5x + x^2$$ □

8.4.2 Use Scilab to build the polynomial with roots $1, 2, 3, -1$.

Solution:

```
// Use the poly() function with vector roots
x = poly(0, "x");
p = poly([1 2 3 -1],"x")

--> p = poly([1 2 3 -1],"x")
```

$$p = -6 + 5x + 5x^2 - 5x^3 + x^4$$ □

8.4.3 Find the roots of the polynomial $x^3 + x^2 + x + 1$.

Solution:

```
p = [1 1 1 1];// x^3+x^2+x+1 =0
roots(p)
--> roots(p)
```

```
ans  =

    4.441D-16 + 1.i
    4.441D-16 - 1.i
   -1.        + 0.i
```

The roots are $-1, i, -i$. □

8.5 Homework: Applications and Differential Equations

8.5.1 Compute the following integrals using Sci lab integration functions (intg, int2d, int3d):

1.

$$\int_2^3 x^3 \sin(x^2)\,dx$$

8.5.2 Find the solutions of the equation $e^{-x^2} + 0.1(x-1)^2 - 0.5 = 0$ by using the Scilab command fsolve. Hint to choose an initial guess, graph the equation and visually select a good starting guess.

8.5.3 Compute the integrals:

1. $\int_0^{2\pi} e^{\cos x}\,dx$
2. $\int_0^1 \sqrt{x^3 + 4}\,dx$

8.5.4 Solve the first order differential equation $\dfrac{dy}{dx} = \dfrac{y}{x} + x\cos x$ on the interval $[1,25]$ with initial condition $y(1) = 0$

8.5.5 Plot the solution of the forced motion

$$x'' + 4x' + 4x = \cos 2t$$

with initial condition $x(0) = 1, x'(0) = -2$ over the interval $[0,20]$ and draw the phase plane with vector field.

8.5.6 Solve the second order differential equation $\dfrac{dx^2}{dt^2} + 25x = \sin(5.5t)$ on the interval $[0,30]$ and with initial condition $x(0) = 0, x'(0) = 0$. Hint rewrite the equation as a system of equation that is

$$\frac{dy_1}{dt} = y_2; \qquad \frac{dy_2}{dt} = -25y_1 + \sin(5.5t)$$

and use the pendulum script to plot the phase plane and the solution on different figures.

Solution:

```
function [ydot]=rside(t,y);..
    yd1=y(2);yd2=-25*y(1) +sin( 5.5*t);..
    ydot=[yd1; yd2 ];
    endfunction

subplot(2,1,1)
t=0:.02:30;
x0=[0 ; 0]; t0=0;
[sol]=ode(x0,t0,t,rside);
plot(sol(1,:), sol(2,:), 'r' )
subplot(2,1,2)
plot(t,sol(1,:))
```

$\square$

8.5.7 Solve the second order differential equation $\dfrac{dx^2}{dt^2} + 0.5\dfrac{dx}{dt} + 25x = \sin(5t)$ on the interval $[0,30]$ and with initial condition $x(0) = 1, x'(0) = 1$. Plot the phase plane and the solution on different figures. Hint for the right side:

```
function [ydot]=rside(t,y);
    yd1=y(2);yd2=-.5*y(2) -25*y(1)+sin( 5*t);
    ydot=[yd1; yd2 ];
    endfunction
```

8.5.8 Given the two polynomial functions:

$$f(x) = -4x^2 + 2x + 3$$

$$g(x) = x^4 - 4$$

Answer the following
1. Plot the two functions and find their intersection points.
2. Find the area of the region between the two functions interior to the two intersection points.
3. Find the area of the region formed by $f(x)$ above the x-axis.
4. Find the area of the region formed by $g(x)$ below the x-axis.

8.5.9 Solve the first order differential equation $e^x y' - y^2 = x$ on the interval $[0,10]$ with initial condition $y(0) = 2$

8.5.10 Plot the solution of the second order differential equation

$$x'' + 25x = \sin 5.4t$$

with initial condition $x(0) = 0, x'(0) = 0$ over the interval $[0,30]$ and draw the phase plane. The displayed solution is an example of the beat phenomenon.